T0361813

Geographical Information Systems and Spatial Optimization

Geographical Information Systems and Spatial Optimization

Sami Faiz
Associate Professor in Computer Science
Faculté des Sciences Juridiques
Economiques et de Gestion
Jendouba
Tunisia

Saoussen Krichen
Associate Professor in Quantitative Methods
Faculté des Sciences Juridiques
Economiques et de Gestion
Jendouba
Tunisia

CRC Press
Taylor & Francis Group
Boca Raton   London   New York

CRC Press is an imprint of the
Taylor & Francis Group, an **informa** business

A SCIENCE PUBLISHERS BOOK

CRC Press
Taylor & Francis Group
6000 Broken Sound Parkway NW, Suite 300
Boca Raton, FL 33487-2742

© 2013 Copyright reserved
CRC Press is an imprint of Taylor & Francis Group, an Informa business

No claim to original U.S. Government works

Printed in the United States of America on acid-free paper

International Standard Book Number: 978-1-4665-7747-3 (Hardback)

This book contains information obtained from authentic and highly regarded sources. Reasonable efforts have been made to publish reliable data and information, but the author and publisher cannot assume responsibility for the validity of all materials or the consequences of their use. The authors and publishers have attempted to trace the copyright holders of all material reproduced in this publication and apologize to copyright holders if permission to publish in this form has not been obtained. If any copyright material has not been acknowledged please write and let us know so we may rectify in any future reprint.

Except as permitted under U.S. Copyright Law, no part of this book may be reprinted, reproduced, transmitted, or utilized in any form by any electronic, mechanical, or other means, now known or hereafter invented, including photocopying, microfilming, and recording, or in any information storage or retrieval system, without written permission from the publishers.

For permission to photocopy or use material electronically from this work, please access www.copyright.com (http://www.copyright.com/) or contact the Copyright Clearance Center, Inc. (CCC), 222 Rosewood Drive, Danvers, MA 01923, 978-750-8400. CCC is a not-for-profit organization that provides licenses and registration for a variety of users. For organizations that have been granted a photocopy license by the CCC, a separate system of payment has been arranged.

Trademark Notice: Product or corporate names may be trademarks or registered trademarks, and are used only for identification and explanation without intent to infringe.

```
              Library of Congress Cataloging-in-Publication Data
Faiz, Sami.
   Geographical information systems and spatial optimization / Sami
Faiz,
Saoussen Krichen.
        pages cm
   Includes bibliographical references and index.
   ISBN 978-1-4665-7747-3 (hardback)
   1.   Geographic information systems. 2.   Geospatial data.   I. Ti-
tle.
   G70.212.F35 2013
   910.285--dc23
                                                          2012036756
```

Visit the Taylor & Francis Web site at
http://www.taylorandfrancis.com

CRC Press Web site at
http://www.crcpress.com

Science Publishers Web site at
http://www.scipub.net

Foreword

This is an important work at a time when economic, social and political success depends on ever-greater efficiencies in the use of financial, human and natural resources. A popular cell phone may say on the label that it is made in China, but this is just the surface of the story. The company whose plant assembles the phones in Guangzhou is part of a technology firm based in Taiwan. Circuit boards used in the phone are made in Malaysia from parts sourced via Singapore. Critical chips are sent from Europe, the USA, S. Korea and elsewhere. The cases that house these sleek phones are fabricated in Indonesia and the specially treated glass that serves as an interactive screen comes from plants in North America. The software that makes the finished product useful comes from someplace far distant from the final assembly, or where it is eventually used. This means that each of these devices and indeed all of our latest information age hardware are the amalgamation of globe circling supply chains. A recent analysis presented in The Economist (January 21st 2010, page 84) maintains that the Chinese portion of the production of the popular iPad accounts for only 2% of the cost, even though the label says, "Made in China".

One can readily appreciate the fact that the more efficiently all these supply chains are organized, the lower the overall costs and consequently the greater the chance of success in the market place. It is no wonder that all spheres of society from agriculture, to better government to the environment are all in a state of rapid evolution using the techniques highlighted in this book. In modern agriculture it is commonplace for the most advanced machinery to be connected by GPS to land information databases so that just the right mix of inputs are monitored to avoid waste and maximize output. Indeed the global food supply, food security as well as famine relief all depend on the application of the models outlined in this work in order to be sure that food arrives where it is most needed and in sufficient quantity and quality. Large commercial food brokers such as Cargill as well as international food aid agencies like the United Nations World Food Program constantly monitor the location, routing and capacities of virtually every suitable vessel on the high seas in order to keep the world food system running on the one hand and to meet unanticipated food shortage emergencies, on the other. Thus, the problem of optimal packing and optimizing routing discussed by the authors is a 24/7 activity that, for better or worse, we all depend.

Everywhere government budgets are incapable of providing all the services demanded by their citizens. One of the major causes of the fact that budgets just don't seem to go as far as in the past is something called Baumol's Cost Disease.[1] William J. Baumol and his colleague William G. Bowen observed that some parts of society such as the performing arts were not able to benefit from the cost savings of modern technology, as did other parts of the economy. The cost of public security, education, healthcare, environmental protection and social welfare continue to climb, putting ever greater strains on finite public capacity. The major point for our purposes here is that public institutions have been left holding the bag of these parts of the economy that are the least amenable to productivity gains.

If public services are to be available, then the means of providing them must be optimized. We can see how some of these services have so productively responded to the opportunities covered in this work that they have been taken over by the global economy. Just one example is the combination of GIS databases with optimization models in parcel and package handling. This was traditionally a government provided service and national postal services evolved into powerful near monopolies. In recent years, companies such as DHL, UPS and others have so reinvented the entire way in which items are sent, processed and delivered from one point to another that they have largely supplemented the old government mail service for anything that is important. Thus, the quest for good governance is just as dependent on finding ways to apply the lessons covered by professors Faiz and Krichen in this book.

---

[1] Baumol, William J. and William G. Bowen (1996), Performing Arts: The Economic Dilemma, New York: The Twentieth Century Fund

The major contribution of these authors is that they provide a way of going beyond thinking about problems of logistics to how we might solve them.

Dr. Jim Riddell
International Center for Land Policy Studies
and Training, Taiwan

Contents

List of Tables

List of Figures

List of Algorithms

Introduction

Geographical information systems (GIS) are becoming powerful environments for managing large volumes of data in potential frameworks such as transportation, network management and urban planning. The main characterization of GIS is the ability of describing each entity by spatial data through its location as well as through shape.

GIS offer various operations as capturing, storing, managing, analyzing and displaying geographically referenced data. They are developed by using database management systems (DBMS) structured in such a way as to facilitate the extraction of useful information and the acquisition of knowledge for decision making.

For instance, many complex situations require not only a simple exploration among data, but also a generation of the best solution. It is from this viewpoint that the integration of GIS and optimization tools can interestingly generate profitable alternatives. Such a framework, "GIS-O", is designed to provide needful data that on the one hand supply optimization routines and on the other hand display the generated solution within a mapping format.

Subsequently, to find out the most profitable solution, optimization routines become unavoidable and to interpret better strategic decisions, GIS turn to be compulsory.

Among the GIS's functionalities, we can point out the shortest path problem and the choice of a subset of placement nodes from a large set of nodes. Such problems are known to be exponentially growing with the problem size. To proceed to an efficient response to such queries, the solution approach, as illustrated in figure 1, follows a first step that consists in highlighting the required view in terms of the specified layers. The resulting view and its numerical data are transferred to optimization tools that consists in modeling the problem and computing its complexity, then specifying a suitable solution approach compromising in terms of quality and computational time. The outcome is finally displayed in a cartographic format.

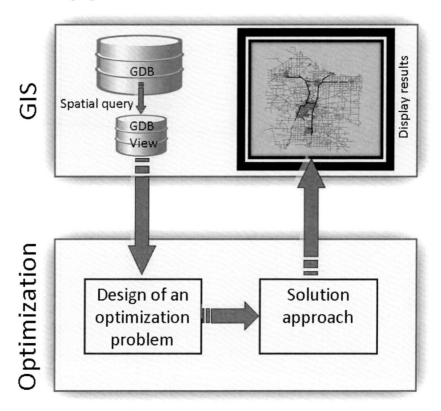

Figure 1 The process of decision making inside a GIS.

Color image of this figure appears in the color plate section at the end of the book.

In this book, we'will expose basic concepts of GIS and optimization separately. Then, we will demonstrate through numerous related applications, their capabilities in managing the operations. However, some limits arise while designing such environments.

To remedy these drawbacks, we propose to integrate these environments and overview the existing literature. It follows after a detailed survey that three main classes of integration strategies are of interest (Bivand and Lucas, 1997 and Malczewski, 2006). These integration strategies were applied to numerous fields of study as hydrology and water resource (Olivera and Maidement, 1999;

Sui and Maggio, 1999 and Schumann et al., 2000), environment and ecology (Groenigen et al., 1996; Marulli and Mallarach, 2004 and Rossi and Villa, 2009), waste management (MacDonald, 1996; Sumathi et al., 2007 and Alvarez et al., 2007) and urban planning (Gomes and Lins 2002; Wang et al., 2004 and Li, 2011). Other topics addressed the routing problems that, in most cases, necessitate the dynamic display of data and the generation of satisfactory solutions. Therefore, an integration of GIS and optimization tools can provide more adequate solutions.

As a potential application, we consider the class of routing problems that encompasses the vehicle routing problem (VRP) and its variants. Among those variants, we can refer to a tractable problem, known as the two-dimensional capacitated vehicle routing problem (2L-CVRP), previously studied by Zachariadis et al. (2009) and Duhamel et al. (2011).

This book investigates the vector loading capacitated vehicle routing problem with distance constraints (VL-DCVRP) that consists in loading vehicles with orders, to be delivered to geographically dispersed customers and designing the optimal route for each vehicle. We started by stating mathematically the addressed problem, then a specific integration strategy will be proposed involving both of the GIS and optimization tools to obtain a more suitable framework for solving such routing problems.

This book is organized as follows:

- **Chapter 1** enumerates the main concepts of GIS that encompass geographical databases (GDB) while stating their main variations. Among these GDB, we explore spatio-temporal databases that store the historical states of already accomplished actions.

- **Chapter 2** introduces the main concepts of optimization that contribute in generating profitable solutions. A detailed statement of the optimization problem's main components with the definition of the optimality and efficiency are also identified. Some practical optimization problems and solution approaches, namely exact and approximate methods, are also surveyed.

- **Chapter 3** overviews GIS-O integration strategies split up into three main classes: the full, the loose and the tight protocols. A comparative study is also pointed out to show the effectiveness of each strategy. Real applications are exposed for illustrating the use of either strategy. It is noticeable, based on this survey, that the use of a GIS-O integration strategy is application-dependant.

- **Chapter 4** is devoted to a loose GIS-O integration approach for solving the vector loading capacitated vehicle routing problem with distance constraints. The new environment is very comfortable in solving and visualizing the routing solution.

Chapter 1

Geographical Information Systems: Basic Concepts

1.1 Introduction

1.2 Geographical databases

1.3 Geographical information systems

1.4 Research areas

1.5 Conclusion

1.1 Introduction

In the past, maps served as the most prominent tool for illustrating and developing informed decisions by using spatial information. Now, Geographical Information Systems (GIS) are available in almost all institutions dealing with spatial study. The management of rapidly growing volumes of data, was solved by using high configuration machines. The main factor that makes this possible is the storage of geographical entities over time.

There is a problem in handling geographical objects in their current state as well as in their historical account. As a result, spatio-temporal databases were designed and proposed as alternative representations to make the available data more informative and to offer the possibility of simulation or prediction of some strategic events.

This chapter is an introduction to geographical and spatio-temporal databases. At a more general level, the book highlights the need of a GIS in various fields of study. In fact, various applications can perform better when managed by a GIS thanks to the large panoply of functionalities and a convivial environment displaying the obtained solutions in a clear way.

This chapter is organized as follows: Section 2 deals with geographical and spatio-temporal databases followed by their main features and alternative representations. Section 3 outlines the main concepts of a geographical information system. Section 4 is devoted to the research areas within geographical databases (GDB) and GIS.

1.2 Geographical databases

In numerous decision situations, one is required to locate strategic placements or compute shortest paths, once a source-destination couple is specified. To do so, maps are generally used to show in a clearer way the solution details and discuss its impact from different points of view. Historically, maps have been the most powerful and practical tool for recording data, designing ideas and projects, communicating and taking rational decisions. Therefore, available information within maps are indexed by their spatial and descriptive data (Tomlinson, 1984; Dangermond, 1988):

- **Spatial data:** For the location and the shape of the map components.
- **Descriptive or factual data:** Reporting the features of objects as the name, the height and the population.

In GDB, the relationship between objects is multidimensional. As an illustration, positioning vocabularies are to be performed as *"near, far from, before, after, in the left hand, in the right hand, in, adjacent to"*. Graph theory can be a promising design of such positions while describing the adjacency, and its cost, for every pair of locations in the map.

1.2.1 Features of geographical data

Numerical geographical data are mainly characterized by (Faiz, 1999):

- **Large volume:** Geographical data are voluminous therefore they encompass numerous layers and themes.
- **Abstract data type:** Spatial information does not follow a specific type, as integer or character, but a generic representation of data as points, lines and polygons.
- **Multiple sources of data:** Geographical data come from different and heterogeneous sources. They are measured through captors or on the territory using a positioning aid system of type "Global Positioning System" (GPS), aerial photos and satellite images. Geographical data can also be tracked from maps or existing plans (Laurini and Thompson, 1992 and Laurini and Milleret-Raffort, 1993).
- **Multi-scale character:** Only distinguishable objects can be plotted easily. Based on such details, the scale determines the existence of an object and its pertinence in the framework.
- **Complex encoding:** The same geometrical form can define more than one spatial object (a line can alternatively correspond to a

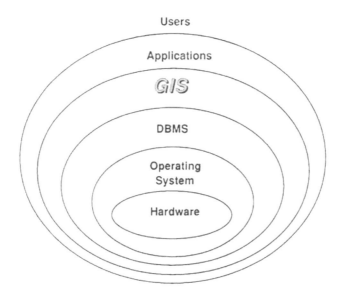

Figure 1.1 Global GIS framework.

road or a river). Conversely, a given spatial object can be modeled by alternative geometrical forms (a city can be represented by a point or a polygon according to the adopted scale). Thus, geographical maps are clearly designed in terms of the addressed context.

- **Temporal aspect:** States of objects can change over time. An object can be created or deleted. As long as it exists, an object can change its semantic attributes or its geometrical form or position. For example, a vehicle can move from one position to another and an agricultural zone can become urban. In this vein and depending on the subject, some users can be tempted to preserve a historical sample of data.

1.2.2 Representation of geographical data

Geographical databases are characterized by descriptive and spatial information. The spatial component can be designed regarding two main modes, namely:

- *Vector-based representation:* Used to record basic geographical coordinates as isolated points (trees or mountain summits), lines (roads or rail networks) and polygons (parcels or cities). Vector-based data are encoded directly through the digitalization of lines located on 19 the map, as reported in figure 1.2. The related algorithms for managing such data are mainly inspired from manual experience. The major advantages of this encoding are related to numerical data and spatial analysis by the use of the topology between objects (intersection, adjacency or overlay).

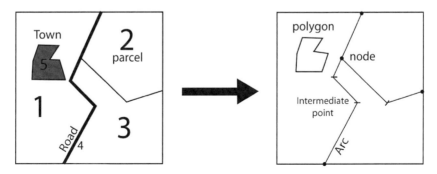

Figure 1.2 From the map to the vector-based representation.

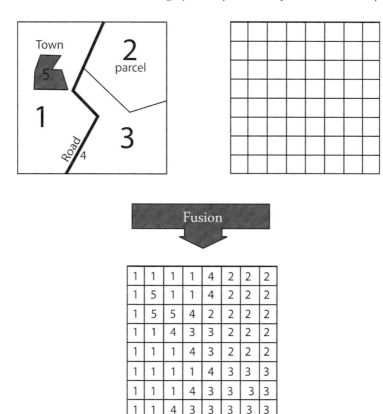

Figure 1.3 From the map to the cell-based representation.

Meanwhile, the duration, the cost of the input and the difficulties encountered in the case of overlapping information layers reveal some limitations for such a representation.

- *Cell-based representation*: Also called matrix representation. It can be obtained using a data entry form or a meshing on a basic cartographic document as shown in figure 1.3. The meshing corresponds to a scale and a data projection. The resolution of the final representation depends on the size of mesh. Aerial photos and numeric satellite images require generally a cell-based encoding. This representation is specified by its fastness and cost saving of the data inputs. However, it should be pointed out that there is a lack of geometrical accuracy and semantic poverty. Subsequently, it is impossible to manage the topological operations.

For the vector-based representation, the information is interpreted in the input phase. However, for the cell-based representation, the information is stored in its initial form. By adopting the vector-based representation, a subset of points, lines and polygons are to be drawn in order to describe the objects' delimitations. For instance, in the cell-based mode, a set of pixels is stored. Thus, the vector-based information is better structured and well adapted to querying operations. However, the cell-based representation is more informative by displaying the area under study as an image.

To sum up the above comparison, we propose the encoding of the land by the use of a vector-based representation for strategic objects and/or a cell-based representation for objects to be displayed. The best design of an application is to select, according to the context, the suitable representation: a decision maker in a phone network who aims to determine the shortest path for reaching a new city and seeks to know accurately a location of the nearest phone room, will obviously perform vector-based data. Meanwhile, a pedologist asked to schedule cuts in terms of the ground and the vegetation pattern will be tempted by a cell-based representation. A relevant extension for geographical databases consists in adding the temporal dimension to the stored data. Such representation is characterized by numerous functionalities stated in the next section.

1.2.3 Spatio-temporal databases

Spatio-temporal databases include objects that can change in terms of their coordinates or shapes during a period of time (Faria et al., 1998; Ryu and Ahn, 2001 and Koubarakis et al., 2003). In such cases, acquisition techniques- as the GPS- are to be performed to locate mobile objects and deliver these coordinates to the server to be updated or stored. As an illustration of spatio-temporal objects, let us consider a car on a highway. A large variety of applications use such objects for which their positions and coordinates are recorded along time. Generally, traditional databases assume that data is stored statically. A modification can be made only after a manual update. It is more suitable to design a model that predicts the future movement of objects based on the historical data. An explicit function having a temporal aspect can be developed to be performed later for predicting future positions and behavior of objects. Taking into account such temporal aspects, the management of spatio-temporal databases requires efficient access approaches defined as spatio-temporal queries (Xie and Shibasaki, 2005).

A spatial predicate is defined as a point or a shape whereas a predicate involves a fixed time or a time interval. We can point out various types of spatial and temporal queries:

1. **Selection query:** Two sub-types of such queries arise:
    (a) Timestamp or Timeslice query: It consists in finding all objects inside a specific zone at a given time.
    (b) Range query: It generates all objects inside a specific zone during a time interval.
2. **Neighborhood query:** Identify the nearest object from a reference point during a time interval.
3. **Total query:** Find the number of objects inside a specific area within a time interval.
4. **Juncture query:** Find the number of aircrafts that will travel within 20 km in the next 15 minutes.
5. **Similarity query:** Given a specific zone, enumerate the instants where 12 objects are inside.

1.3 Geographical information systems

According to some definitions, a GIS is a powerful set of tools for acquisition, storage, management, retrieval, analysis and display of geographical data (Laurini and Thompson, 1992). The GIS can be seen as an interactive information system capable of handling geographic phenomena (Chrisman, 1999; Cowen, 1988; Faiz, 1999 and Korte, 2000). Various terms can designate a GIS as:

• Reference spatial information systems
• Locating system information
• Spatial DBMS
• Geographical DBMS

1.3.1 Historical note

During the past decades, systems for automatic processing of spatial data (or GIS) have evolved. The first GIS was developed in Canada in the 1960s, called the "Canadian Geographical Information System" (Tomlinson, 1984) and proposed by the Ministry of agriculture. It's purpose was to make a census of the agricultural land in Canada. This project involved the creation of maps highlighting the urgent rehabilitation of agricultural land and thus allowed the proper distribution of subsidies. During the sixties, there were several initiatives for the development of GIS in universities particularly SYMAP ODYSSEY of Harvard university, GEOMAP of university of Waterloo and MANS of university of Maryland (Dangermond, 1982).

Techniques for converting Raster to vector and vice versa were developed while maintaining the link between data pointers and Database Management Systems (DBMS).

More recently, other techniques were mastered as there are no effective methods to convert a large number of maps in digital formats due to the small storage capacities of computers and their slow running. The most efficient machine available was the IBM 1401 with 16 K-bytes of memory. It performed about 1 000 instructions per second. In April 1964, IBM 360/65 greatly contributed in the evolution of GIS, with a memory of 512K-bytes and 400 000 instructions per second. At that time, the magnetic tape was the most used storage material.

Henceforth, computers became a master tool for meteorologists, geologists and geophysicists. For instance, there is a lack in the interactivity and graphical editing capabilities due to the expensive cost of the graphical displays.

In the early 1970s, teams of GIS development took place around the world. The first conference on GIS was held in 1970 and was attended by 40 participants. The second was in 1972 with 300 participants. Both meetings were held in Ottawa. More recently, the number of conferences grew significantly. A first course in automated cartography was planned in universities and business agencies. Henceforth, commercial geographical applications were developed.

Among the most known companies, we refer to ESRI (environmental systems research institute) and Inter-graph.

A rapid advance in electronics had enabled the design of computers with large memory sizes, faster, more interactive and less expensive. Programs dealing with spatial data became faster but missed spatial analysis.

In the early 1980s, the falling of prices and the increasing power of computers were an incentive for the apparition of a large number of teams operating in the field of GIS as the management of large volumes of data and the display of complex maps became easier. The application of GIS has increased. Scholars began to feel the need to identify the data that has the greatest impact on decision making.

The cost, availability and size of computers have improved. Query processing has been optimized and the responses have been edited in real time. However, in some cases, the computer device was unable to manage the large volume of stored data.

Nowadays, progress-especially in the database- combined with graphic functions, enable GIS to provide a framework to achieve much better spatial analysis (complex operations in less time and the opportunity to handle larger volumes of data). Up to now, there has been a considerable gap between the technology used in geographical analysis and the number of users applying it. Subsequently, a clear and accurate information on GIS should be provided as an incentive to involve more users. When developing a language of spatial information, it is worthwhile to keep in mind previously developed methods designed for manual mapping as a reference.

1.3.2 Typology of GIS

We can classify GIS in terms of the nature of the managed data. Such systems generally belong to the following categories:

- *GIS for land administration*: Various queries can be of interest as the land enhancement, the identification of owners, cadastral plans, urban development and regional integration projects, solar study, plans and models of relief development, planning and management of land use, surface computing and the management of green spaces and parks (Cornia and Riddell, 2008).

- *GIS for statistical mapping*: Maps surveying statistical results, demographics maps, socio-economic, epidemiological, tourism, maps and management utilities, asset management of cities, natural hazard mapping and risk of type accidents, crimes or out-of-pipe, mapping simulation of propagation of fire, sliding, flooding, mapping and management of natural resources such as forest management, agriculture or environmental and biological quality of air and water, geomarketing (implementation study of a new facility or agency, network design display advertising, as well as newspapers and regional daily newspapers).
- *GIS for network management*: Road maps, bus routes and other similar services for vehicles, rail and river networks, distribution of drinking water, gas, electricity and telephone. The spatio-temporal database generates plans, with respect to alternative scales strongly correlated to different areas, to control the situation of the network and reflects its extensions. It also provides practical assistance in the daily management of networks and facilitates the planning and preparation of strategic tasks. The geomatic application allows the storage and display of relevant events and the consideration of statistics and development of decision support systems.
- *GIS for remote sensing images*: Geodetic survey, impact on the environment, mineral exploration, coastal processes, study of ocean pollution, cultures and archeology.

Numerous geographical phenomena can be faced in real life situations as: local authorities (municipalities and regional councils), the major ministries (environment, agriculture, equipment and housing, transportation, finance and defence), operators of natural resources (agriculture, fisheries, petroleum, and mines), meteorology, the offices of the topography and cartography, national geographical institutes, publishers of atlases, firms surveyors, distribution companies (water, electricity and gas), transport companies (buses, taxis and trains), the automotive manufacturers that incorporate GIS in vehicle design (integration of a navigation system and guidance for map display), companies of posts and telecommunications, construction companies, large organisms (equipment, planning and civil security), offices of broadcasting (network implementation for radio and television transmissions), marketing services (set up subsidiaries and plan tours for commercial agents), banks, agencies handling socio-economic data, statistics, educational institutions and research.

1.3.3 Potential applications

As mentioned above, a GIS is a decision aid tool that can be applied in many ways to many problems. We can print out some potential applications as:

- **Transportation:** Detection of high traffic to re-organize group travels and determine optimal paths. Such studies can permit us to conclude about relationships between people belonging to the same cities. In this context, a GIS for DHL (a Deutsche shipping company) was developed.
  Its activity is to deliver, using 1 400 vehicles, goods over geographically dispersed customers. The major concern was to sort items in terms of the postal code and deliver, with minimal cost, the goods using some of the available vehicles. Two main concerns are to be avoided when analyzing the situation: the waste of time and the cost. To remedy to the problem of sorting and delivery, DHL express opted for ESRI GIS using ArcGIS Server software. The use of the GIS environment improved the whole process in both time and cost saving. Tasks are handled automatically using the soft by firstly sorting items in terms of their postal codes, then displayed in the map while highlighting the routes to be adopted.
- **Political activities:** Location of voters and designing a cartography for displaying the final results. A relevant application of GIS to voting and elections offers a reliable source of information about the area under study, as the finding and analysis of demographic and behavioral features of the population (Shyy et al., 2007). GIS is also able to provide information about the repartition of the political parties and propose eventual locations for the polling. To strengthen the analysis, GIS offers the possibility of displaying the above mentioned information in the map.
- **Water resources:** Study of the network scaling to satisfy consumers' demands.

- **Public lighting:** Location of luminous points and electrical networks.
- **Waste management:** Determination of the suitable number of containers, depots and optimal paths for waste collecting.
- **Environment:** Flood areas, nuisance areas, propagation of aerial and aquatic pollution.
- **Security:** Delimitation of fire bounds and simulation of natural disasters.
- **Urbanism:** Land use and development plans.

Regarding the above examples, GIS are viewed as efficient tools for recording, exploiting, analyzing and displaying geographical data.

1.3.4 Main components of a GIS

A GIS is an automatic system that merges data from various sources. It collects, organizes, stores, manages, analyzes and visualizes information geographically localized to enrich the knowledge of the addressed space. As outlined in figure 1.4, a GIS contains three main components:

1. **Data acquisition**
   - Acquisition of large volumes of data in many different ways: digitalizing maps, scanning images, remote sensing, aerial photography, GPS, terrain and existing files.

2. **Processing**
   - Storage, handling, operation and spatial analysis.
   - Interrogation and treatment (command language, spatial query language, interactive manipulation with icons and menus).
   - Sets (intersection, inclusion, union).
   - Geometry (calculating distances, perimeters and areas).
   - Conversion (scale, projection, internal representation: vector to raster and vice versa).
   - Overlay of layers.
   - Networks as optimal path finding.

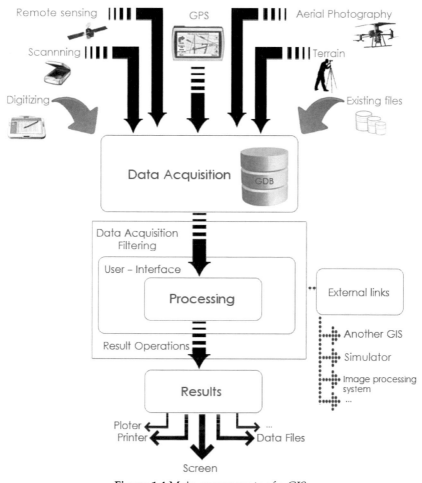

Figure 1.4 Main components of a GIS.

Color image of this figure appears in the color plate section at the end of the book.

- Selection (value or location).
- Spatial analysis (generation of polygon, generation of adjacent elements and generation of buffer zones).
- Transformation (rotation and translation).
- Map overlay data and image (overlay of thematic layers on a single division of geographical space, overlay different thematic divisions, superimposition of vector data and raster image).

- Implementation of spatial relationships between geographical entities (intersection, inclusion, difference, inside, disjoint, meet and adjacent to).
- Spatial aggregation (common geometrical elements, the sum of an attribute weighted by the surface or the length of each element).
- Statistics (generation of histograms, pie charts, definition of class limits, calculation of average, standard deviation, variance, sorting on variables).
- Measures (counting objects by class).
- Linking the GIS with external devices as an image processing system or a simulator.

3. **Results**
- Multimedia processing (integration of images in digital form).
- Visualizing the obtained results in three dimensions (cross-sectional views, perspective views).
- Windowing and zooming.
- Publishing and visualizing spatial data in tabular formats, charts or histograms.
- Choosing the scale and the projection.
- Creating interactively graphical symbols and legends.

At any level, the GIS can necessitate to interact with external environments. For the data acquisition, the GIS can import data from a scanner, a GPS or other existing databases. In the processing step, external links are generally required as the interaction with another GIS or the questioning of large DBMS and Data files.

1.3.5 GIS functionalities

We can identify different types of operations to be performed by a GIS. Among the available functionalities in such environment, we can point out the search of data in terms of a date, an address or a path. Other queries can be managed by the GIS as the finding of a suitable path given the departure and arrival coordinates (or address) and the computation of some statistical metrics as the normal and peaks. From an economic standpoint, the main issue concerns the best user satisfaction and the minimization of the water loss from reservoirs to consumers. For the construction or expansion of any kind of network, land morphology, parcels or any other situation as congested areas and free roads, the immediate availability of spatio-temporal databases saves time.

Using these operations, the GIS is able to answer user's queries as:

1. **Factual**
   - Give the numbers and the addresses of persons owning parcels.
   - Show the distribution of the population which own parcels by city.
   - What are the names of all highways?
2. **Graphical**
   - View the parcels belonging to persons outside the city.
   - Show the parcel located at 29 Africa Avenue.
   - Display regions of Tunisia where wheat is grown.
3. **Spatial**
   - The parcels located within 100 meters of a fire hydrant.
   - Pathways that lead from Tunis to Kairouan (see figure 1.5).
   - The perimeter of the block number 125 (assuming that such information is not stored).
   - Specific locations where electricity network crosses the water distribution network.
   - What are the wadis that cross the city of Jendouba?
   - The cities of over 110 000 inhabitants belonging to the designated area on the screen.
   - View parcels in front of 29 Africa Avenue.
   - Draw in green the country with an area exceeding 420 000 square kilometers whose growth rate is above 1.6 and in red countries whose area is less than 230 000 square kilometers whose growth rate is below 1.1%.
   - The largest delegation in the region of Kairouan.
   - Overlaying the map of administrative areas with that of the land use.

The first three queries are called evidence queries (standard treatment and standard output). They can be solved by using any traditional DBMS. The fourth, fifth and sixth requests are called

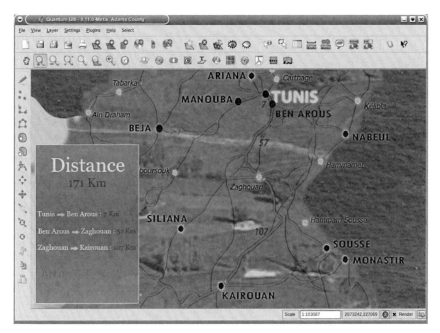

Figure 1.5 Optimal pathway from Tunis to Kairouan.
Color image of this figure appears in the color plate section at the end of the book.

graphics (standard treatment and graphical output). The resulting outputs can be issued as a graph, displaying geographical entities according to specific criteria. The last ten queries are called spatial (particular treatment and standard or graphical output). This type of queries involve the spatial aspect of data: the topological relations between geographical entities. The query results can be graphic or not. For this reason, current DBMSs cannot handle such queries in a simple way.

1.3.6 Advantages and drawbacks of GIS

The main benefits of GIS are:

- Storage of large volumes of geographical data at a low cost.
- A lower cost of producing maps and plans.
- The generation of rapid mapping with an interactive spatial selection process.

- The establishment of new maps and plans that cannot be achieved manually.
- The improvement product presentation and brand owner.
- The quick access to geographical data.
- The update and analysis of diverse geographical data.
- The study of changes between two or more dates.
- The update in real time makes the GIS a tool for monitoring.
- Gathering, in the same system, data from different sources and combine them.
- The integration of new requirements with a minimum cost.
- The reduction of uncertainty.
- Effective planning and management.
- Speed and efficiency in decision making.
- Increased productivity.
- Improving services rendered to the user.

The main drawbacks of GIS are:

- Lack of specialized personnel and competent implementation of a GIS project.
- High cost and technical problems during data acquisition.
- Cost of maintenance and administration of up-to-date data.
- Cost of software maintenance of the product and the lack of flexibility in programming.
- Non-standardization of data types and spatial query languages.
- Interaction of different types of data and different databases.
- Lack of tools for integrating the dynamic visualization of geographical information in three dimensions (3D).
- Lack of tools for time management.
- The quality of information processed and the lack of tools for its management are irrelevant.

1.4 Research areas

The digital geographical information is still a recent technique. For instance, many problems encountered by the decision maker (DM) remain unsolved. Therefore, a real research program should be established involving all stockholders (users, data producers, researchers, academics, business software and services). These stockholders will be involved in different research areas:

- Modeling of geographical data.
- Standardization of formats for the exchange of digital map data.
- Management of temporal variations and combinations of information from different scales.
- Methods and algorithmic generalization of digital map information.
- Real time updates.
- Automation of mapping update by using satellite imagery.
- Automatic generation of maps.
- Knowledge acquisition from spatial databases using data mining techniques.
- Multidimensional representation of spatial data for a quick and efficient interpretation and decision making using spatial OLAP.
- Metadata documentation and representation.
- Quality control.
- Certification of geographical databases.
- Full inter operability of systems, data and networks.
- Combination of GIS with different tools such as remote sensing systems for project development or simulation.
- Improved performance: query optimization and benchmarking.
- GeoWeb that corresponds to the process of design, implementation, production and delivery of maps on the Web.
- Analysis and interpretation of satellite imagery using inference mechanisms integrated in DBMS.
- Graphical user interface (GUI) and manipulation languages adapted to the handling and processing of geographical hyper-media data.
- Dynamic visualization and visual simulation for impact studies (planning, development, environment).

As such, the following techniques can be of great relevance:

1. **Databases**
    - Design pattern (hyper-media data models).
    - Access methods for spatial data (spatial indexes).
    - Integration of inference mechanisms (deduction).
    - Databases spread (distribution of data on different sites).

- Graphical user interface and extensive manipulation languages (spatial query).

2. **Artificial intelligence**
   - Mechanisms of inference (expert systems).
   - Pattern recognition.
   - Connectional models and neural networks.
   - Multi-agent systems.

3. **Programming languages**
   - Object-oriented language.
   - Graphical user interfaces (GUI).
   - Parallelism.

4. **The computer architecture**
   - Networks of servers and workstations.

5. **Optimization**
   - Generation of efficient solutions to the problems of loading and routing.
   - Finding ways and channels of lower costs in complex networks.
   - Capacity optimization of antennas in telecommunications networks.
   - Fragmentation and package data in networks.

1.5 Conclusion

We surveyed, in the present chapter, the main concepts related to a GIS with spatial and spatio-temporal dimensions. To be more realistic, some practical situations were detailed in order to show the importance of GIS in managing such a huge volume of data. We also highlighted the benefits and limits of the existing GIS as an incentive for the integration of the optimization techniques inside GIS to be more efficient in the finding of best solutions. To do so, we need to introduce, optimization, its basic concepts and definitions in both single and multiple objectives frameworks.

Chapter 2

Optimization: Basic Concepts

2.1 Introduction
2.2 Design of an optimization problem
2.3 Features of an optimization problem
2.4 Potential problems in optimization
2.5 Solution approaches
2.6 Conclusion

2.1 Introduction

Decision making is a strategic action that requires optimization techniques to be accomplished while considering resource constraints. The optimization can be viewed as a powerful tool, expressed as a quantitative modeling of the problem that, once solved, outputs the best solution as a rational decision that corresponds to the most profitable one. It starts by the definition of the main goal of the decision maker (DM). As this goal should take into account structural requirements, a set of constraints are to be enumerated. Thus, the obtained model is an alternative specification of the addressed problem, named optimization problem. Depending on the problem's features, as the complexity, a solution approach can be adopted to generate optimal or near optimal solutions. The main steps that characterize an optimization problem are firstly, the specification of the objective(s) to be optimized (minimized or maximized) in terms of the decision variables. Then, the statement of feasibility constraints.

We can note that two main classes of optimization problems arise:

- *Single objective optimization problems*: Including only one objective to be optimized. Its resolution yields to the optimal solution that corresponds to the best value of the addressed objective. If the number of decision variables does not exceed 3, the graphical resolution can be a practical way to generate the optimal solution. However, for a larger number of decision variables, various solution approaches, depending on the problem structure, were proposed in the literature. For the linear case, the appropriate method is the simplex algorithm (Dantzig, 1956).
- *Multiobjective optimization problems*: Considering a set of objectives to be concurrently optimized. Under this framework, the resolution yields to a set of efficient (or Pareto optimal) solutions that constitute the frontier of the feasible space.

  If a decision problem disposes of p objectives expressed in terms of n decision variables, the dimensionality of the objective space is p and the dimensionality of the decision space is n. The set of efficient solutions is plotted in the objective space forming an efficient frontier to be determined analytically using the suitable solution approach.

Applications performing optimization techniques are countless. Among the challenging problems, we can cite the knapsack (Martello and Toth, 1988), the bin packing (Krichen and Dahmani, 2010), the transportation, telecommunication networks (Krichen et al., 2012a), the assignment and the scheduling.

We outline, in the present chapter, the main concepts that define and characterize an optimization problem in both single and multiobjective frameworks. Then, we state the main steps to be adopted to identify, formulate and solve a decision problem. Such a statement is then illustrated by the description of some optimization problems followed by their mathematical models. As the resolution of such problems can be exact or heuristic, we detail the main principles of each class and underline the conditions for their use.

The remainder of this chapter is split up into six main sections. Section 2 states the basic components of an optimization problem. Section 3 describes the optimization problem's features followed by some fundamental definitions. Section 4 reports some potential optimization problems illustrated by their mathematical formulations and their importance in modeling some practical situations. Section 5 addresses two main solution approaches for solving an optimization problem namely exact approximate methods.

2.2 Design of an optimization problem

An optimization problem is described through some generic key elements that should be adapted to the problem under study. The main elements reported in figure 2.1 are:

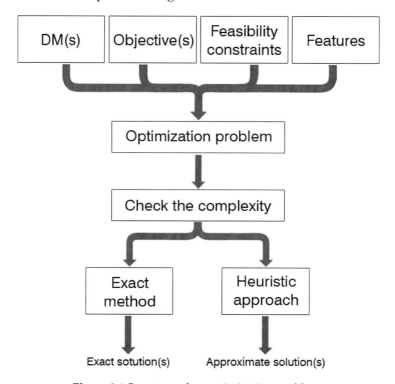

Figure 2.1 Structure of an optimization problem.

- **Decision makers DM(s):** The number of DMs involved in the optimization process followed by their objectives.
- **Objective(s):** Depending on the number of objectives to be taken into account, the problem can be placed in the:
  - Single objective framework that generates only one optimal solution.

- ▪ Multiobjective framework that generates a set of efficient solutions.
- **Feasibility constraints:** Define the area to be explored in order to locate the best solution(s).
- **Features:** Some additional qualifications for the problem under study can be of great interest for its characterization. The problem can be:

  - ▪ Linear or non linear.
  - ▪ Deterministic or stochastic.
  - ▪ Static or dynamic.
  - ▪ Combinatorial or with continuous decision variables.

A DM is required to take a rational decision in which case an objective (or goal) is already defined and numerically expressed. Generally speaking, the DM should respect structural constraints when designing his decision. An optimization problem follows generally the mathematical model (2.1):

$$Max\ Z(x)$$

$$S.t.\ D = \{gi(x) \le 0, i = 1,\dots, p\} \qquad (2.1)$$

To be well formulated, an optimization problem needs the definition of some basic terms. We enumerate in what follows the most useful optimization vocabulary:

- *The objective function* is an expression of the DM's preferences to be maximized.
- *The constraints* are equalities/inequalities to be taken into account when maximizing or minimizing the objective function.
- *The feasible space D* corresponds to region delimited by constraints $gi(x) \le 0, i = 1,\dots, p$ expressed in the formulation (2.1). Each solution inside *D* is called a feasible solution. The main purpose when solving problem (2.1) is to find x^* located inside *D*.
- *Decision variables* are the unknown values of the optimization problem. Solving the decision problem yields to finding these values in such a way to maximize the objective function and fulfill the set of constraints.

- *The solution* is a vector $x = (x_1,..., x_n)$ of n decision variables that express the numerical value of each decision variable.
  $x^* = (x_1^*,..., x_n^*)$ denotes the best solution of the problem (2.1).

Such a situation is called an optimization problem (OP).

Table 2.1 illustrates the design of an optimization problem. The objective function $Z(x)$ is an explicit expression of the DM's goal that can be either maximized or minimized. A set of p constraints, defining the feasible space, are to be respected. The objective function and the set of constraints are expressed in terms of these decision variables in order to be computed while being in the feasible space.

Table 2.1 Main components of an optimization problem.

	Objective function	**Constraints**
Notation	$Z(x)$	$gi(x)\ i = 1,..., p$
Nature	A straight line	A set of straight lines
Cardinality	1	p

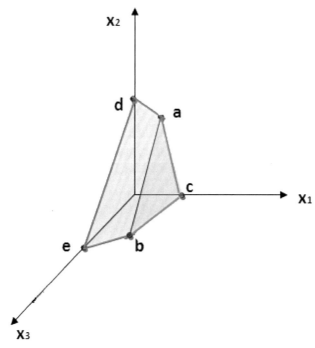

Figure 2.2 Feasible space of an optimization problem.

Color image of this figure appears in the color plate section at the end of the book.

The feasible space is delimited by the constraints. The problem consists in finding the solution that maximizes the objective function.

Hence:
- **Inputs** of an optimization problem are the equations enumerated in the mathematical model (2.1).
- **Outputs** are the values of the decision variables. These values constitute the optimal solution for the problem (2.1).

To clarify the needs of an optimization problem in practical situations, we need to survey its main features followed by some basic definitions.

2.3 Features of an optimization problem

The main characterization of an OP is the number of objectives to be considered by the DM. Two main classes are depicted namely the single objective and the multiobjective cases. The definition of a solution depends strongly on this classification. In fact, for the single objective case, the resolution of the optimization problem yields to an optimal solution that corresponds to the best objective function's value while re-specting structural constraints. Thus, two different definitions for the best solution follow, depending on the number of objectives. We can point out, as reported in figure 2.3 that in the single objective case (2.3 (a)), only one optimal solution follows. However, in the multiobjective framework (2.3 (b)), a set of efficient solutions is

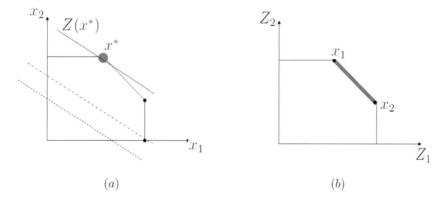

(a) (b)

Figure 2.3 Graphical representation of an optimization problem.

an alternative concept for the optimality. We develop the optimizing in each care in the remaining part of this section.

2.3.1 Single objective optimization

When an optimization problem includes only one objective function, each solution is evaluated according to the objective as a scalar. For the single objective case, the best solution is called "optimal" (Dantzig, 1956).

Definition 1 *A solution x^* is said optimal if its objective value $Z(x^*)$ corresponds to the maximum (with respect to problem (2.1)) value while respecting the set of structural constraints.*

This definition is illustrated in figure 2.3 (a) by x^*. For illustration, we can refer the reader to section (2.4.1) in which a numerical example is developed and the optimal solution is generated using the CPLEX software, one of the most used solvers.

An optimization problem can be either linear, in which case both the objective function and the constraints are linear. To solve a linear optimization problem, the simplex method is generally applied and the optimal solution is obtained within a finite number of iterations. If it is assumed that all decision variables are discrete, the branch and bound method is the suitable approach for generating the optimal solution. For the nonlinear case, the problem becomes more complex and its resolution requires other approaches that generally provide near optimal solutions.

An example

Let us consider a numerical example defined mathematically as follows:

$$Max\ Z(x) = 3x_1 + 2x_2$$
$$S.t.\ 2x_1 + x_2 \le 30 \qquad (2.2)$$
$$x_1 - 2x_2 \le 10$$
$$x_1, x_2 \ge 0$$

Where the solution is described by the vector $x = (x_1, x_2)$ of decision variables. We show graphically in figure 2.4 the optimal solution x^* that corresponds to the best value of $Z(x)$ while being in the feasible space.

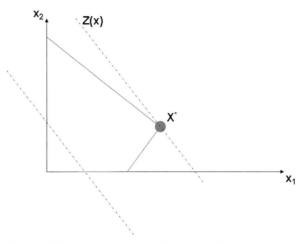

Figure 2.4 Graphical representation of the optimization problem (2.2).

2.3.2 Multiobjective optimization

When a set of p objectives are to be optimized concurrently, the optimization problem is called *multiobjective* (Ulungu and Teghem, 1994). The following formulation accounts for a multiobjective problem:

$Max\ Z_1(x)$

$Max\ Z_2(x)$
$\vdots$

$Max\ Z_p(x)$

$S.t.\ D = \{gi(x) \le 0,\ i = 1,..., p\}$ (2.3)

A feasible solution x is evaluated according to each objective. Hence, a p-dimensional objective vector is expressed as:

$$Z(x) = \begin{pmatrix} Z_1(x) \\ Z_2(x) \\ \vdots \\ Z_p(x) \end{pmatrix}$$

For illustration, if we consider $p = 2$ objectives and assume that two solutions x_1 and x_2 are evaluated according to their respective objective vectors $Z(x_1)$ and $Z(x_2)$:

$$Z(x_1) = \begin{pmatrix} 100 \\ 20 \end{pmatrix} \qquad Z(x_2) = \begin{pmatrix} 80 \\ 40 \end{pmatrix}$$

The main proposal is the one to choose between these two solutions. Neither solution is to be discarded as x_1 is the best according to the first objective and x_2 is the best according to the second objective.

Hence, if we consider a solution x_3 evaluated as $Z(x_3) = \begin{pmatrix} 50 \\ 50 \end{pmatrix}$, it should be accepted as a promising solution. Following this vectorial comparison, x_1, x_2 and x_3 are not comparable and should be retained.

Subsequently, the optimality is replaced by the concept of efficiency that gives rise to a set of efficient solutions. We state below the definitions of both efficiency and set of efficient solutions.

Definition 2 The efficiency

A solution x is said to be efficient if there does not exist any other solution y in the feasible space such that $Z_j(y) \geq Z_j(x)$, $\forall j = 1,...., p$ (Ulungu and Tehgem, 1994).

Definition 3 The set of efficient solutions

The set of efficient solutions ES corresponds to the solutions that cannot be comparable with respect to all objectives, that is:

$$ES = \{x \in D \mid x \text{ is efficient}\}$$

As the number of objectives p ($p \geq 2$) gives rise to a vector of evaluations, numerous solutions can be non comparable with respect to the Pareto principle. For example, if two solutions x_1 and x_2 are evaluated according to two criteria to be maximized as below:

$$Z(x_1) = \begin{pmatrix} 30 \\ 50 \end{pmatrix} \text{ and } Z(x_2) = \begin{pmatrix} 70 \\ 30 \end{pmatrix}$$

Both x_1 and x_2 belong to the Pareto set ($x_1, x_2 \in ES$).

An example

As an illustration for the multiobjective framework, let's consider the following formulation (Krichen et al., 2012b and Masri et al., 2012):

$$
\begin{aligned}
Max \quad & x_1 \\
& x_2 \\
& x_3 \\
S.t. \quad & 2x_1 + 3x_2 + 4x_3 \leq 12 \\
& 4x_1 + x_2 + x_3 \leq 8 \\
& x_1, x_2, x_3 \geq 0 \qquad\qquad (2.4)
\end{aligned}
$$

Where the number of objectives is 3, the number of decision variables is 3 and the number of constraints is also 3. This example, plotted in figure 2.2, corresponds to a three-dimensional representation of the feasible space where the coordinates of solution are expressed in terms of the decision variables x_1, x_2 and x_3. We highlight in figure 2.2 the efficient set of solutions in red color. The set ES is bounded by the points $\{a, b, c\}$.

To sum up the above features, we report the following decomposition in table 2.2, that classifies the optimization problems in four types depending on:

- *The number of objectives*: Single objective/multiobjective.
- *The linearity*: Linear/non linear.

Table 2.2 Classification of optimization problems.

	Linear	Non linear
Single objective	Optimality	Optimality
	Simplex method	Golden sections
Multiobjective	Efficiency	Efficiency
	Multiobjective simplex method	Metaheuristics

Based on the above description and characterization, we survey in the rest of this chapter some well and widely studied optimization problems known to be *NP*-hard.

2.4 Potential problems in optimization

We overview in this section some relevant optimization problems largely studied in the literature, namely:

- The knapsack problem (KP)
- The bin packing problem (BPP)
- The container loading problem (CLP)
- The assignment problem (AP)
- The scheduling problem (SP)
- The traveling salesman problem (TSP)
- The capacitated vehicle routing problem (CVRP)

Due to their importance in practical situations, we start by stating each problem, then enumerating some variants and related potential applications.

2.4.1 The knapsack problem

We detail in what follows the knapsack problem (KP), characterized by its ability in modeling any other decision situation, due to its simplicity and generic features. To understand more clearly the mathematical concept of the problem, an illustrative example is stated, modeled and solved using the CPLEX optimizer. The KP consists in considering a set of n items from which a subset is to be stowed in a knapsack with a pre-fixed capacity W. Each item $i = 1,..., n$ is characterized by its reward *pi* and its weight *wi*. The selected subset must be the most profitable and should not exceed the capacity of the knapsack. It should be pointed out that both the objective function and the capacity constraint are linear. Hence, KP, in its basic version includes:

- *A single objective*: The maximization of items to be loaded in the knapsack.
- *The capacity constraint*: Loaded objects should not exceed the capacity of the knapsack.
- *The linearity*: Both the objective function and the capacity constraint are linear.
- *The decision variables*: Binary.

Notation

The following notation accounts for the definition of the basic KP:

Symbols	Description
INPUTS	
n	The total number of objects
p_i	The reward of object i, $i = 1, \ldots, n$
w_i	The weight of object i, $i = 1, \ldots, n$
W	The capacity of the knapsack
OUTPUTS	
$x = (x_1, \ldots, x_n)$	The solution in its vectorial representation
$x_i =$	$\begin{cases} 1 & \text{if object } i \text{ is in the knapsack} \\ 0 & \text{elsewhere} \end{cases}$

Mathematical model

The mathematical formulation of the basic KP (Martello and Toth, 1990) is:

$$Max \ Z(x) = \sum_{i=1}^{n} p_i x_i$$

$$S.t. \ \sum_{i=1}^{n} w_i x_i \leq W$$

$$x_i \in \{0, 1\} \quad i = 1, \ldots, n$$

(2.5)

For illustration, let's consider the numerical configuration for a KP with $n = 5$ objects and a capacity of the knapsack $W = 100$ reported in table 2.3. A feasible solution for this problem can be described as follows:

$$x = \boxed{1 \ \ 1 \ \ 1 \ \ 0 \ \ 0}$$

(2.6)

According to the representation (2.6), we can notice that:

1. Objects 1, 2 and 3 are loaded in the knapsack, however objects 4 and 5 are outside.
2. The total weight corresponding to this solution is: 50 + 30 + 20 = 100. Hence, the knapsack is saturated.
3. The solution reward is: $Z(x) = 20 + 3 + 5 = 28$.

Table 2.3 Dataset for a KP with $n = 5$.

Object	Reward	Weight
1	20	50
2	3	30
3	5	20
4	30	40
5	10	70

The mathematical formulation of this problem is stated as follows:

$$Max \ Z(x) = 20x_1 + 3x_2 + 5x_3 + 30x_4 + 10x_5$$

$$S.t. \ 50x_1 + 30x_2 + 20x_3 + 40x_4 + 70x_5 \leq 100$$

$$x_1, \ldots, x_5 \in \{0, 1\}$$

(2.7)

Performing the CPLEX software, the mathematical program (2.7) is solved and the optimal solution is $x^* = 50$. Figure 2.5 corresponds to the mathematical formulation (2.7) written in CPLEX. Figure 2.6 reports the resolution of the model (2.7). The difficulty in solving the KP is proportional to the number of objects n. In fact, it was shown that, for large values of n, exact methods are not able to solve the KP. Subsequently, approximate approaches perform well in generating near optimal solutions in a reasonable computational time (Martello and Toth, 1990).

```
CPLEX> display problem all
Maximize
 obj: 20 x1 + 3 x2 + 5 x3 + 30 x4 + 10 x5
Subject To
 c1: 50 x1 + 30 x2 + 20 x3 + 40 x4 + 70 x5 <= 100
Bounds
 0 <= x1 <= 1
 0 <= x2 <= 1
 0 <= x3 <= 1
 0 <= x4 <= 1
 0 <= x5 <= 1
Binaries
 x1  x2  x3  x4  x5
```

Figure 2.5 Mathematical formulation of a KP with $n = 5$ by CPLEX.

```
MIP - Integer optimal solution:  Objective = 5.0000000000e+001
Solution time =    0.06 sec.  Iterations = 1  Nodes = 0

CPLEX> disp sol var -
Incumbent solution
Variable Name            Solution Value
x1                           1.000000
x4                           1.000000
All other variables in the range 1-5 are 0.
```

Figure 2.6 Solution details for a KP with $n = 5$.

2.4.2 The bin packing problem

One of the most challenging problems in operations research is the class of packing problems that can be seen as variants of the bin packing problem (BPP). The basic version of the BPP consists, given a set of items and a prefixed number of bins, in packing the whole set in a minimum number of bins without violating capacity constraints (Loh et al., 2008). Various practical situations can be modeled as a BPP, namely, the maritime transportation of goods, the packing problem with revenue management and the clustering of items with homogeneity constraints.

Notation

We enumerate in what follows the main symbols used in the mathematical model (2.8)–(2.12) of the BPP:

Symbols	Description
INPUTS	
n	The total number of items to be packed
p	The total number of available bins
w_i	The weight of item i
W	The capacity of each bin
OUTPUTS	
$x_{ij} = \begin{cases} 1 & \text{if item } i \text{ is loaded in the bin } j \\ 0 & \text{elsewhere} \end{cases}$	
$y_j = \begin{cases} 1 & \text{if the bin } j \text{ is used} \\ 0 & \text{elsewhere} \end{cases}$	
$x = (x_{ij}, y_j)$	The vector of decision variables

Mathematical model

The objective considered in the basic version of the BPP is to minimize the number of used bins among the set of p bins. Kantorovich (1960) proposed the following mathematical formulation:

$$Min \quad Z(x) = \sum_{j=1}^{p} y_j \tag{2.8}$$

$$S.t. \quad \sum_{i=1}^{n} w_i x_{ij} \leq W y_j, \ j = 1, \ldots p \tag{2.9}$$

$$\sum_{j=1}^{p} x_{ij} = 1, \ i = 1 \ldots n \tag{2.10}$$

$$y_j \in \{0, 1\}, \ j = 1, \ldots, p \tag{2.11}$$

$$x_{ij} \in \{0, 1\}, \ i = 1, \ldots, n \text{ and } j = 1, \ldots, p \tag{2.12}$$

Where:

- **The objective function** (2.8) consists in minimizing the total number of used bins.
- **Constraints**
  - Inequalities (2.9) denote that the total weight of items packed in a bin should not exceed the capacity W of that bin.
  - The second set of constraints (2.10) require each item to be packed in exactly one bin.
  - Binary variables are defined in equations (2.11) and (2.12). A solution $x = (x_{ij}, y_j)$, $i = 1,..., n$, $j = 1,..., p$ reports the used bins as well as the packing configuration of the items in bins.

Numerous variants of the BPP were studied in the literature as the online BPP. Another variation consists in assuming that the bins are sized differently. The dimensionality of bins can also give rise to additional extensions of the BPP when evaluating each item according to its length and width. Hence, the problem becomes a two-dimensional BPP that requires more structural constraints and makes its resolution more complex.

2.4.3 The container loading problem

In the area of production and distribution of cargos, the efficient use of transportation devices as containers and palettes is of extreme economic relevance. A prohibitive consumption of the applied transportation capacities produces a significant cost saving, reduces the goods traffic and protects the natural resources.

The two-level packing of items, in containers then in compartments, is called the container loading problem (CLP). The optimization of the CLP yields to the most profitable solution that corresponds to the minimum sum of remaining spaces in the used containers.

Such a problem arises in various large-scaled industrial applications and transportation networks. Even with the importance of the cargo loading operation in the actual logistics, packing freights into containers is still manual in many warehouses.

As the problem is *NP*-hard, the alternative use of exact or heuristic approaches is required regarding the problem size.

A satisfactory solution for the CLP can belong to the economic scaling as well as it can take into account an ecological standpoint due to the adverse consequences of high traffic on environmental resources.

The study of some variations of the CLP, as in the dynamic case, shows the cost saving spread over many periods, which is successfully accomplished using commercial softwares and solvers. Other variations of the CLP can include the multiple CLP (Che et al., 2011), where various types of containers are considered for the packing of items. As the cost depends on the type of container, the objective of the DM for this variation is to minimize the total packing cost.

The CLP can alternatively be seen in a three-dimensional perspective while minimizing the remaining space in the used containers (which is equivalent to the minimum number of containers).

Besides, Egeblad et al. (2009) studied the strip packing version of the CLP. Some characteristics of the CLP are of interest as the dimensionality, the cargo assortment and the number of objectives.

- **The dimensionality** is a geometrical characteristic that can handle one, two, or three dimensions.
  The basic version of the CLP is one-dimensional in which case the objective is to maximize the total weight of containers. Scheithauer (1992) studied this aspect and showed that when the CLP is one-dimensional, it can be modeled as a KP.
  The two-dimensional CLP is a generalization of the first case in the sense that it involves the length and width of objects. A solution of the two-dimensional CLP describes how to stow objects in containers. A survey of this topic is presented by Lodi et al. (2002).

The three-dimensional CLP adds to the two-dimensional version the consideration of the height whenever items are evaluated.

- **The cargo assortment** consists in packing items having the same form and dimensions. In this context, Dyckhoff (1990) and Wascher et al. (2007) proposed the following classification:

  - Homogenous cargos: Cargos are denoted homogenous if all items have the same dimension(s).
  - Heterogeneous cargos: Two main levels of heterogeneity can be pointed out:
    - *Weakly heterogeneous*: If the items can be grouped into relatively few classes according to their sizes and shapes.
    - *Strongly heterogeneous*: If only a very few number of items can be grouped into classes.

- **The number of objectives** is of great relevance for the CLP as it determines the concept of solution to be adopted.

  - The single objective case: Only one of the previous objectives was considered (Khanafer et al., 2010).
  - The multiobjective case:
    - The biobjective case (Liu et al., 2008 and Krichen and Dahmani, 2010).
    - More than two objectives (Imai et al., 2006).
      The following are some objectives addressed in the literature as the minimization of the number of containers Zhu et al. (2011), the minimization of the containers' unused volumes, the maximization of items' weights and the maximization of the total priority.

Among the above variations, the single objective one-dimensional CLP corresponds to the less complex version. As it is about how to find the minimum number of containers required to pack the items without violating capacity constraints (Loh et al., 2008), many approaches were developed to reach promising solutions in a reasonable running time.

Notation

Symbols	Description
INPUTS	
n	Number of items to be shipped
w_i	Weight of item i
p_i	Priority of item i
m	Number of containers on the ground
W	Capacity of each container
n_{comp}	Number of compartments
m_{max}	Maximum capacity of the cargo hold
m_a	Weight of the aircraft before loading
m_{max}^k	Maximum weight of cargo that can be loaded in compartment k ($k = 1, \ldots, n_{comp}$)
OUTPUTS	
x_{ijk}	0-1 variables that indicate if item i is placed in bin j and compartment k.
y_{jk}	0-1 variables that determine if bin j is assigned to compartment k ($y_{jk} = 1$), or not ($y_{jk} = 0$)

Mathematical model

We state thereafter a biobjective version of the CLP proposed by Krichen and Dahmani (2010), where each item is characterized by its weight and priority:

$$Max \quad Z_1(x) = \sum_{i=1}^{n} \sum_{j=1}^{m} \sum_{k=1}^{n_{comp}} w_i x_{ijk} \qquad (2.13)$$

$$Max \quad Z_2(x) = \sum_{i=1}^{n} \sum_{j=1}^{m} \sum_{k=1}^{n_{comp}} p_i x_{ijk} \qquad (2.14)$$

$$S.t. \quad \sum_{i=1}^{n} w_i x_{ijk} \leq W y_{jk}, \; j = 1, \ldots, m, \; k = 1, \ldots, n_{comp} \qquad (2.15)$$

$$\sum_{j=1}^{m} \sum_{k=1}^{n_{comp}} x_{ijk} = 1, \ i = 1, \ldots, n \tag{2.16}$$

$$\sum_{k=1}^{n_{comp}} y_{jk} = 1, \quad j = 1, \ldots, m \tag{2.17}$$

$$\sum_{i=1}^{n} \sum_{j=1}^{m} \sum_{k=1}^{n_{comp}} w_i \, x_{ijk} \leq m_{max} \tag{2.18}$$

$$\sum_{i=1}^{n} \sum_{j=1}^{m} w_i x_{ijk} \leq m_{max}^{k}, \ k = 1, \ldots, n_{comp} \tag{2.19}$$

$$y_{jk} \in \{0, 1\}, \ j = 1, \ldots, m, \ k = 1, \ldots, n_{comp} \tag{2.20}$$

$$x_{ijk} \in \{0, 1\}, \ i = 1, \ldots, n, \ j = 1, \ldots, m, \ k = 1, \ldots, n_{comp} \tag{2.21}$$

Where $x = (x_{ijk}, \ y_{jk})$, $i = 1, \ldots, n$, $j = 1, \ldots, m$ and $k = 1, \ldots, n_{comp}$.

Objectives

- The first objective (2.13) seeks to maximize the total weight of the load.
- The second objective (2.14) tries to maximize the total priority of the load.

Constraints

- The first set of constraints (2.15) mean that the capacity of a container should be respected for each items' combination. It is noticeable that the right side of the inequality obliges y_{jk} to take the value 1, when an item i is assigned to container j.
- Constraints (2.16) impose that each item is packed in exactly one container which is packed in exactly one compartment.
- Constraints (2.17) ensure that each container j must be loaded at most once.
- Inequalities (2.18) and (2.19) express the maximum weight capacity constraints for each compartment.
- The decision variables are to be binary as required in equations (2.20) and (2.21).

2.4.4 The assignment problem

An assignment deals with the problem of associating elements to others belonging to two different sets while taking into account some problem requirements. The assignment problem (AP) is one of the extensively studied optimization problems in the literature of operations research because of its ability in modeling a large variety of real applications. A relevant case study related to the traffic assignment is due to Zhang et al. (2010) that aims to minimize the traffic cost.

The basic version of the AP consists in linking elements belonging to two separate sets while trying to minimize the linking cost.

Assigning task i to agent j requires a cost c_{ij}. Therefore, the objective within an AP is to depict the configuration of pairs (i, j) that corresponds to the minimal total cost while fulfilling system constraints (Nemhauser and Wolsey, 1988).

Notation

The following notation accounts for the mathematical model described in equations (2.22)–(2.25):

Symbols	Description
INPUTS	
n	The total number of tasks
m	The total number of agents
c_{ij}	The cost of production when assigning task i to agent j
OUTPUTS	
x_{ij}	0-1 variables that indicate if task i is assigned to agent j

Mathematical model

The basic version of the AP is modeled as follows:

$$Min \quad Z(x) = \sum_{i=1}^{n} \sum_{j=1}^{m} c_{ij} x_{ij} \tag{2.22}$$

$$S.t. \quad \sum_{j=1}^{m} x_{ij} = 1, \ i = 1, \ldots n \tag{2.23}$$

$$\sum_{i=1}^{n} x_{ij} \le 1, \ j = 1, \ldots, m \tag{2.24}$$

$$x_{ij} \in \{0, 1\}, \ i = 1, \ldots, n, \quad j = 1, \ldots, m \tag{2.25}$$

- **The objective function** (2.22) is to minimize the total assignment cost of tasks to agents.
- **Constraints**
  - Constraints (2.23) require that each agent should accomplish exactly one task.
  - Constraints (2.24) indicate that each task has to be realized by exactly one agent.
  - In constraints (2.25) all decision variables are to be binary.

2.4.5 The scheduling problem

An alternative optimization problem of great relevance, named the scheduling problem (SP), is defined as the sequencing of a set of tasks onto periods while taking into account temporal precedence requirements (Esquirol and Lopez, 1999). The SP seeks, in most cases, to minimize the total completion time of activities, called the makespan.

The SP has been largely studied in the literature in numerous versions. Basically, the problem involves a set of tasks to be scheduled and can require the use of some resources for their accomplishment. The solution is an ordering between those tasks with regard to both system constraints and resources availabilities. For more details about the main features that can characterize a SP, we can refer the reader to Blazewicz et al. (2001).

A main variation of the SP, called the resource constrained project scheduling problem (RCPSP), is stated as follows: given a project described by a set of activities V numbered from 0 to $n + 1$, each activity i has to be processed without preemption. Some assumptions to accomplish the project are outlined:

- **Dummy activities:** The consideration of two additional activities, called dummy activities, $i = 0$ and $i = n + 1$, correspond to the starting and the termination of the project, where their respective durations are $d_0 = d_{n+1} = 0$.
- **Resources availability:** K renewable resources are considered in the project. The availability of each resource k in each time period is R_k ($k = 1,...,K$) units.
- **Resources duration:** Each activity i requires r_{ik} units of resource k during each period of its duration where $r_{0k} = r_n + _{1k} = 0, k = 1..., K$ showing that dummy activities do not require any resource).
- **Precedence relations:** For each activity i, a set of predecessors is to be defined. Thus, a set P enumerates all couples (i,j) such that i is one of the predecessors of j.
- **Solution representation:** A solution to the RCPSP is a sequence of starting times $x = (s_0, s_1,..., s_{n+1})$, with $s_0 = 0$.

Notation

The following notation accounts for the mathematical model of the RCPSP expressed in (2.26)–(2.30):

Symbols	Description
INPUTS	
V	The set of activities
K	The set of ressources
S	The set of precedence relations
n	The total number of activities
P	The set of predecessors
d_i	The duration of activity i
r_k	The units available for each resource k
r_{ik}	The units of resource k required by activity i
OUTPUTS	
s_i	The starting time of activity i

Mathematical model

Formally, the RCPSP can be stated as follows (Kolisch and Padman, 2001):

$$Min \quad Z(x) = Max_{i=1}^{n}\{s_i + d_i\} \tag{2.26}$$

$$S.t. \quad s_i + d_i \leq d_j, \ (i,j) \in S_j \tag{2.27}$$

$$\sum_{i \in A(t)} r_{im} \leq r_m, \ m = 1, \ldots, k, \ t \geq 0 \tag{2.28}$$

$$s_0 = 0 \tag{2.29}$$

$$s_i \ integer, \ \forall i = 1, \ldots, v \tag{2.30}$$

- **The objective function** (2.26), also denoted by C_{max}, is the minimization of the project makespan.
- **Constraints**
  - Precedence constraints (2.27) force activity j not to be started before each of its predecessors i where precedence relations are expressed in E.
  - Constraints (2.28) define at each time t the total resource demand as less than or equal to the resource availability for each resource k, where $A(t)$ refers to the set of activities in progress at time t, $A(t) = \{i \in V | s_i \leq t < s_i + d_i\}$.
  - Constraint (2.29) imposes the project to begin at time instance zero.
  - Constraints (2.30) ensure that the activity start times are non negative integers.

2.4.6 The traveling salesman problem

The traveling salesman problem (TSP) is about finding the best pathway to visit n cities, so that each city is visited once, then return to the starting point. Such a cycle is called a Hamiltonian circuit (Moon et al., 2002). The main objective in the TSP is to generate the Hamiltonian cycle that corresponds to the minimal cost, or distance.

The TSP has long been studied in the literature due to its applicability in many fields as sequencing problems, manufacturing processes or routing optimization. The TSP, in its simplest version, can be seen as a graph $G = (V, E)$ in which the set V of vertices represents cities and E, the set of edges, expresses all possible routes between cities. The TSP is defined through the following assumptions:

- The main concern of the DM is to minimize the total travel distance.
- For each edge $\{i, j\} \in E$ corresponds a travel distance d_{ij}.
- Based on the graph representation, for every pair of nodes $\{i,j\} \in V \times V$, a distance δ_{ij} is computed so that δ_{ij}

$$= \begin{cases} d_{ij} & \text{if } \{i, j\} \in E \\ \infty & \text{elsewhere} \end{cases}$$

- If $d_{ij} \neq d_{ji}$, the TSP is called asymmetric.
- A solution to the TSP corresponds to the Hamiltonian circuit with a minimum distance.

Notation

The following notation accounts for the mathematical model (2.31)–(2.35) of the TSP:

Symbols	Description
INPUTS	
V	The set cities
n	The total number of cities
d_{ij}	The direct route, if exists, between cities i and j
δ_{ij}	The distance between cities every i and j in the graph
OUTPUTS	
x_{ij}	0-1 variables showing if city j is visited immediately after city i

Mathematical model

We develop hereby the mathematical formulation of the TSP (Laporte, 1992):

$$Min \quad Z(x) = \sum_{i=1}^{n} \sum_{j=1}^{n} \delta_{ij} x_{ij} \tag{2.31}$$

$$\text{S.t} \quad \sum_{j=1}^{n} x_{ij} = 1, \ i = 1, \ldots, n \tag{2.32}$$

$$\sum_{i=1}^{n} x_{ij} = 1, \ j = 1, \ldots, n \tag{2.33}$$

$$\sum_{i,j \in S} x_{ij} \geq |s| - 1, \ s \subseteq V, 2 \leq |S| \leq n - 2 \tag{2.34}$$

$$x_{ij} \in \{0, 1\}, \ i, j = 1, \ldots n \text{ and } i \neq j \tag{2.35}$$

- **The objective function** (2.31) tries to minimize the total distance traveled by the salesman.
- **Constraints**
  - The two first sets of constraints (2.32) and (2.33) express that each city must be visited exactly once.
  - Constraints (2.34) prohibit solutions composed of disjoint sub-tours. They are also called constraints for elimination of sub-rounds.

Variants the TSP arise by adding different types of constraints, as visiting each city within a time window.

2.4.7 The capacitated vehicle routing problem

The capacitated vehicle routing problem (CVRP) studied by Vigo (1996), belongs to the class of optimization problems of great relevance due to its usefulness in various fields as supply chain, networks information management and shipping (Ai and Kachitvichyanukul, 2009; Cordeau et al., 2002 and Vigo, 1996).

Formally, the CVRP includes a set of customers whose demands cannot be split into vehicles. Vehicles are assumed to be identical with a fixed capacity and based at a central depot.

According to Toth and Vigo (2003), the largest problems that contain about 50 customers can be consistently solved by exact algorithms. Over this number of customers, the CVRP can only be solved by approximate approaches.

Basically, the addressed objective is the minimization of the total travel cost. For this reason, it would be preferable to use GIS tools to allow a global visualization of the studied area. The CVRP that belongs to the class of difficult combinatorial optimization problems (Marinakis et al., 2010) can be described as follows:

- Goods are to be delivered to a set of customers by a fleet of vehicles from a central depot.
- The locations of the depot and the customers are inputs for the problem.
- The solution is about determining a set of least cost vehicle routes such that:
  - Each customer is served by exactly one vehicle.
  - Each route starts and ends at the depot.
  - Customers' orders assigned to any vehicle must not exceed its capacity.
  - The total length of a vehicle's route must not exceed its distance range.

Notation

Following is the notation adopted in the formulation (2.36)–(2.44):

Notation	Explanation
INPUTS	
m	The number of customers
n	The number of vehicles
W	The capacity of each vehicle
q_i	The order of customer i
d^k	The maximum allowed travel distance by vehicle k
c_{ij}^k	The travel cost between customers i and j by vehicle k
d_{ij}^k	The travel distance between customers i and j by vehicle k
OUTPUTS	
$x_{ij}^k =$	$\begin{cases} 1 & \text{if vehicle } k \text{ travels from customer } i \text{ to } j \\ 0 & \text{otherwise} \end{cases}$

Mathematical model

The CVRP consists in designing a set of vehicle routes for delivering orders to some geographically dispersed customers. The CVRP in its simplest version assumes that all vehicles have the same capacity W and are required to serve all customers with a minimum cost.

Let $G = (V,E)$ be a graph where:

V = {0,1,...,*m*} is the set of vertices representing the depot, indexed by 0, and customers, indexed by 1,...,*m*.

E = {{*i, j*} | *i, j* ∈ *V*} enumerates the availability of direct routes between vertices *i* and *j* of *V*.

Each customer's order must be assigned to exactly one of the k vehicles and the total size of deliveries for customers assigned to each vehicle must not exceed the vehicle capacity W. Besides, each vehicle can travel until a maximum distance d^k. Various configurations of the problem were proposed while considering different structural constraints.

Among those versions, Bodin et al. (1983) suggested the following mathematical formulation:

$$Min \quad Z(x) = \sum_{k=1}^{n} \sum_{i=0}^{m} \sum_{j=0}^{m} c_{ij}^k x_{ij}^k \tag{2.36}$$

$$S.t. \quad \sum_{k=1}^{n} \sum_{i=0}^{m} x_{ij}^k = 1, \ j = 0, \dots, m \tag{2.37}$$

$$\sum_{k=1}^{n} \sum_{j=0}^{m} x_{ij}^k = 1, \ i = 0, \dots, m \tag{2.38}$$

$$\sum_{i=0}^{m} x_{il}^k - \sum_{j=0}^{m} x_{lj}^m = 0, \ \forall k = 1, \dots, n, l = 0, \dots, m \tag{2.39}$$

$$\sum_{i=0}^{m} \sum_{j=0}^{m} d_{ij}^k x_{ij}^k \leq d^k, \ k = 1, \dots, n \tag{2.40}$$

$$\sum_{j=0}^{m} q_j \left(\sum_{i=0}^{m} x_{ij}^k \right) \leq W, \ k = 1, \dots, n \tag{2.41}$$

$$\sum_{j=0}^{m} x_{0j}^k \leq 1, \ k = 1, \dots, n \tag{2.42}$$

$$\sum_{i=0}^{m} x_{i0}^k \leq 1, \ k = 1, \dots, n \tag{2.43}$$

$$x_{ij}^k \in \{0, 1\}, \ i, j = 0, \dots, m, \ k = 1, \dots, n \tag{2.44}$$

- **The objective function** (2.36) consists in minimizing the total cost of all vehicles.
- **Constraints**
  - Constraints (4.13) and (2.38) ensure that each customer is served exactly once.
  - The set of constraints (2.39) ensure the route continuity.
  - Constraints (2.40) show that the total length of each route has a limit.
  - The total demand of any route must respect the vehicle capacity constraint this is guaranteed by the set (2.41).
  - Constraints (2.42) and (2.44) require that each vehicle is used at most once.

2.5 Solution approaches

An optimization model is mainly characterized by its complexity that shows analytically the tendency of solving the problem in terms of the size. In general, two major categories of optimization algorithms are to be pointed out: exact and heuristic algorithms.

An exact algorithm solves the optimization problem to optimality. While heuristic approaches provide near optimal solutions (Shao and

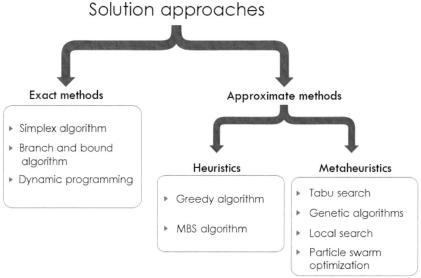

Figure 2.7 Solution approaches for solving optimization problems.

Ehrgott, 2007) in a reasonable computational time. We list below the most famous among exact and heuristic methods and state the main steps to be followed to draw each approach.

2.5.1 Exact methods

An optimization problem can be solved using an exact method to get an optimal solution. If the problem is static and linear in both the objective function and constraints, various solution approaches can be applied as the simplex algorithm. If we add to the problem the combinatorial aspect, the branch and bound approach becomes the suitable method. If the problem is sequential, dynamic programming is adopted. We expose in what follows the main steps for each approach.

1. The simplex method

The most used exact method for solving linear optimization problems is the simplex algorithm (Dantzig, 1956). It proceeds from one breadth-first search or extreme point of the feasible region. It is presented in a tabular form and proceeds iteratively to improve the value of the objective function, moving from one extreme point to another, until reaching the optimal solution.

Many optimization softwares were implemented according to the the simplex method for solving linear mathematical programs as LINDO, CPLEX and GOROBI.

The general guidelines of the simplex algorithm are reported in (10).

Algorithm 1: Main steps of the simplex algorithm

1: Introduce slack variables

2: Provide the initial simplex table

3: **while** (Some marginal values are positive) **do**

4:     Introduce a new variable in the basis

5:     Compute the quantity of this new variable and update the objective
        function value

6: **end while**

7: Display the optimal solution

---

2. The branch and bound algorithm

This approach is performed when decision variables of the OP are discrete. When facing such situation, the feasible set is non convex therefore the simplex algorithm cannot be applied with discrete decision variables. A branch and bound algorithm explores the whole space of solutions in order to generate the optimal solution by a vertical and a horizontal exploration of feasible solutions. Branch and bound algorithms are dependant on the lower and upper bounds. Interestingly, the use of bounds for the function to be optimized combined with the value of the current best solution enables the algorithm to search parts of the solution space and saves time and space according to such behavior. Algorithm (2) is a design of the main steps to be followed for a branch and bound method.

Algorithm 2: The branch and bound method

1: Specify the objective function, lower bounds (LB) and upper bounds (UB)

2: **The branching step:** Split the feasible space D into smaller sets D_1 to D_k

3: **for** $i := 1$ to k **do**

4:    Apply the branching step by finding the optimal solution within D_k

5:    Generate a random number $\beta \in [0, 1]$

6:    Loop recursively to get a branching tree

7:    **The bounding step:** For each generated solution within a subset of the feasible space, determine upper and lower bounds

8:    **The pruning step:** Update UB as the minimum upper bound found. Every node i such that $UL_i > UB$ is discarded

9:    Stop if the feasible space contains only one solution or empty, or if, for some node i, $LB_i = UB_i$

10: **end for**

---

3. The dynamic programming method

It is the process of decomposing the original problem in less dimensional sub-problems and recursively solving it to get the whole solution. Both OPs and computer programming adopted such design of sequential tasks that was originally proposed by Bellman (1953).

Nowadays, there exists a widespread appeal to OPs that are to be solved during a time interval or a prefixed number of periods. In such a context, the problem has to be decomposed into sub-problems connected using backward recursive equations also called optimality equations. Based on Bellman's principle, the optimal solution is obtained by combining solutions of sub-problems according to a backward analysis. Dynamic programming can operate for two main classes of sequential decision problems:

- *Deterministic problems*: Where the move from one state to another is ensured simply by referring to the transition steps given to a current state.
- *Stochastic problems*: The transition from one state to another follows a probability distribution.

We state in algorithm (3) the main steps for developing a dynamic programming approach dedicated to a sequential decision making problem.

Algorithm 3: The dynamic programming algorithm

1: Consider the whole problem and determine the number of periods under study

2: Determine the stages and states of the problem

3: Divide the master problem into D subproblems

4: **for** $i := 1$ to D **do**

5:   Solve each subproblem and determine the solution approach

6:   Apply Bellman's principle to write the optimality equations

7: **end for**

8: Construct the whole solution by combining the partial solutions of the corresponding subproblems starting from the end of the whole process

9: Display the optimal solution or strategy

2.5.2 Approximate methods

Approximate methods can be classified into two main categories: heuristics and metaheuristics. Heuristic approaches are problem-specific in the sense that they cannot be adapted to other OPs. However, metaheuristics are characterized by a set of principles announced generically to be later instantiated depending on the addressed problem. We survey in the rest of this section some approximate methods that can either be heuristics or metaheuristics.

A heuristic method: the greedy algorithm

This class of approximate algorithms was adapted to solve many combinatorial problems as the KP (Teghem et al., 2000), the AP (Tuyttens et al., 2000) and the CVRP (Duhamel et al., 2011 and Marinakis, 2012).

The main principle of a greedy-based algorithm stated in (4) is to construct the whole solution gradually and iteratively by choosing parts of the whole solution while scheduling one step ahead. To adapt the general greedy algorithm, it is required to start by specifying a metric related to the addressed problem that determines how to decide about next choices. The basic idea of the method is to take a decision only for the next iteration and not to change any backward part of the solution even if it is more appropriate.

Algorithm 4: The greedy algorithm

1: Set the solution of size n as an empty set

2: Consider a problem-dependant metric for evaluating a best choice

3: **for** $i := 1$ to n **do**

4: 　Select the best feasible choice for the next iteration

5: 　Remove this choice from consideration

6: 　Add this choice to the whole solution and keep backward choices unchanged

7: **end for**

8: Display the whole solution

Metaheuristic methods

An alternative class of approximate approaches consists in a set of principles that can be adapted to any OP. Two main processes are to be pointed out. Algorithms by neighborhood, as the local search and the tabu search, and algorithms by population, as genetic algorithms and particle swarm optimization.

1. Local search

This methauristic gives rise to numerous algorithms that operate according to an iterative generation of neighborhoods until a stopping rule. Among the numerous algorithms that belong to this class, we refer to the tabu search, the hill climbing and the simulated annealing.

Given an initial solution, the local search (LS) algorithm explores iteratively an appropriately defined neighborhood of current solutions in order to converge to the most profitable configuration. Once a better solution is found, the algorithm proceeds in the same way until accomplishing a prefixed number of iterations. Another stopping criterion consists in repeatedly applying the above behavior until no improvement of the last recorded solution during a predefined landing.

The main steps that characterize the LS algorithm are outlined in algorithm (5).

Algorithm 5: The local search algorithm

1: Generate an initial solution

2: Evaluate the objective function of the initial solution

3: **while** (the termination condition is not satisfied) **do**

4:     Consider the current solution x_c and evaluate its objective value

5:     Generate the neighborhood of x_c denoted by $N(x_c)$

6:     Select the best solution from $N(x_c)$ to be the current solution for the next iteration

7: **end while**

8: Display the best solution depicted during the running of the algorithm

---

2. Tabu search

One of the most promising approximate approaches stated by Glover (1986) is the tabu search (TS), a metaheuristic based on iteratively improving an initial solution until reaching the termination of the process. The TS algorithm (6) starts by generating an initial solution, randomly or using a specific heuristic. The iterative step, that operates while the termination condition is not satisfied, consists in investigating the neighborhood of current solutions. The search operates in such a way as to avoid cycling in the same area using the tabu list that forbids returning to already visited solutions.

Other operators as the intensification and diversification help to improve the quality of the generated solutions. The main components of TS are defined as follows:

- *Tabu list*: Each current solution should be recorded in the tabu list during some iterations in order to forbid cycling in the same region. The tabu list is characterized by its length and the way of recording solutions.
- *Neighborhood*: The neighborhood of an efficient solution is the set of solutions resulting from a transformation of the current solution. The maximum number of neighbors to be explored is predefined in advance. It is required to select from the neighborhood the most profitable solution that can either correspond to the best objective function or a deteriorated solution in the hope of avoiding local optima.
- *Diversification*: The diversification process is based on the generation of different solutions that lead to visit a region of the feasible space not yet explored.
- *Intensification*: The region in which the current solution is located is explored in a more accurate way, during a prefixed number of iterations.

Algorithm 6: The tabu search algorithm
1: Generate an initial feasible solution
2: Set the tabu search parameters
3: **while** (the termination condition is not satisfied) **do**
4:     Generate the neighborhood $N(x_c)$ of the current solution x_c
5:     Select a new solution from neighborhood that does not belong to the tabu list
6:     Update the tabu list
7:     Apply the diversification and intensification operators after a prefixed number of iterations
8: **end while**
9: Save the best solution

We notice that the TS parameters, namely: the tabu list size, the neighborhood's size and the number of iterations are to be adjusted experimentally.

3. Genetic algorithms

They belong to the class of metaheuristics. The basic concept of genetic algorithms (GAs), due to Holland (1962, 1975), is inspired from the natural genetic evolutionary process.

The GA, as annotated in algorithm (7), proceeds as follows: starting by a random generation of the initial population or using specific heuristics, genetic operators as crossover and mutation are applied within each currently observed population. The use of GA requires at first the choice of an appropriate encoding of the chromosome that corresponds to a solution for the addressed problem. GA iterates while recording best encountered solutions. The stopping criterion can differ from one version to another. The details of the genetic operators so far announced can be spelled out as follows:

Algorithm 7: The genetic algorithm

1: Generate an initial population P_0 of $|P_0|$ chromosomes

2: Initialize the crossover and the mutation probabilities α and $\beta \in [0, 1]$

3: Let $c := 1$ be the iteration number

4: $P_c := P_0$

5: Evaluate each chromosome by calculating its fitness function

6: **while** (the termination condition is not satisfied) **do**

7:     Apply the crossover, with respect to α, by selecting randomly two chromosomes c_1 and c_2 from P_c

8:     Perform the mutation operator based on β

9:     Check the feasibility of the obtained chromosomes

10:     Filtrate the new population P_{c+1}

11:     $c := c + 1$

12: **end while**

13: Save the best solution

- *Encoding of a solution*: A solution is designed as a sequence of chromosomes corresponding to fragments of solutions. The representation of a chromosome is problem-dependant. It can be binary, integer, or a string depending on the interpretation of the corresponding codification.
- *Evaluation*: It is generally a computation of the objective function followed by some penalties that deteriorate unfeasible solutions. This step constitutes a basic component of the GA that might considerably influence its performance.
- *Selection*: The selection is made through a random drawing in the interval [0, 1]. Let p be the drawn random number, the selected individual i is the first one satisfying a strongness of each solution compared to p. Hence, the stronger an individual is, the more chances he has to survive.
- *Crossover*: This operator considers two individuals and cuts their chromosome strings at a random position to produce two new

off-springs composed each of the first part of a parent combined with the last part of the other parent. In many cases, the crossover generates infeasible solutions. Therefore, the obtained solutions are to be adjusted.

- *Mutation*: The mutation operator consists in randomly altering one, two or none of the two offsprings depending on an already generated number $\beta \in [0, 1]$. This operator is applied to introduce new information in the population, it is applied in view of avoiding local extremum. However, the mutation probability should be small enough to guarantee the convergence of the GA.

The related literature reveals that GAs are among the most powerful metaheuristics for solving a large variety of OPs.

4. Particle swarm optimization

Particle swarm optimization (PSO), a population-based evolutionary algorithm, is a bio-inspired method developed by Kennedy and Eberhart (1995) to handle combinatorial OPs. As its name designates, a swarm of birds looking for food, starts by flying without having a particular destination until reaching the best food location.

Analogically, PSO initializes a population of random particles corresponding to potential solutions. Each particle is characterized by its position vector and velocity vector. Firstly, particles fly spontaneously inside the feasible space at a random velocity. The main idea of the PSO is to start from an initial swarm, then improve iteratively particles of the swarm by the use of velocities and previously best recorded particles.

Then, velocities are updated based on the best previous particles experiences and the best previous swarm experience. Hence, the behavior inside a population is a compromise between individual and collective memories. This can be described by a swarm intelligence system in which the share of information among individuals is the basic idea.

The PSO script, as outlined in algorithm (8), is described as follows:

- Consider a n-dimensional search space and a swarm of S particles.
- To each particle p in generation i corresponds two vectors x_p^i and v_p^i expressed as follows:
  1. **The location:** A position vector $x_p^i = (x_1^i,..., x_n^i)$.
  2. **The flying velocity:** A velocity vector $v_p^i = (v_1^i,..., v_n^i)$.
- Particles memorize every reached position vector denoted by $x*_p^i = (x*_1^i, \ldots, x*_n^i)$.

- Particles record the whole best position's fitness until generation i. This global best position is referred to as $x_G^i = (x_{g1}^i, \ldots, x_{gn}^i)$.

For each generation i, a particle p takes one of the following decisions:

- Keeping its strategy of search.
- Coming back to its previous individual best position $x*_p^{i-1}$.
- Coming back to the global best position x_G^{i-1}.

The update of the velocity and position are computed as follows:

1. **The new position:** $x_p^i = x_p^{i-1} + v_p^i$
2. **The new velocity:** $v_p^i = w\, v_p^{i-1} + \lambda_1 \left(b_p^{i-1} - x_p^{i-1}\right)$
   $+ \lambda_2 \left(x*_p^{i-1} - x_p^{i-1}\right)$

Where w, λ_1 and λ_2 are predefined parameters.

Algorithm 8: The PSO algorithm

1: Generate the initial swarm where each particle is sized n

2: For particle x in the swarm compute:

    2.1: Its position vector $x_p^0 = (x_1^0, \ldots, x_n^0)$

    2.2: Its velocity vector $v_p^0 = (v_1^0, \ldots, v_n^0)$

3: **while** (the termination condition is not satisfied) **do**

4:    **for** $s := 1$ to S **do**

5:       Evaluate the fitness of the current solution x_p^i

6:       Select the individual best position $x*_p^i$

7:       Select the global best position x_G^i

8:       Update the current position

9:       Update the current velocity

10:    **end for**

11: **end while**

2.6 Conclusion

We reviewed in this chapter the basic notions of an OP, its main components and some related variations. We then defined in a more accurate way the notion of optimal solution for the single objective framework and the efficiency, also called Pareto optimality, for the multiobjective case. For both classes, some specific OPs were exposed to show the effectiveness of a mathematical definition in either solving them at optimality or approximating the best solution(s). We have shown through this analysis the need of the optimization in improving the solution and saving the computational time. An overview of the existing literature shows that there is a need of borrowing an environment from which data are to be extracted and solutions are to be visualized and interpreted. The next chapter is devoted to integration protocols of optimization techniques and GIS through alternative representations. Such integration is called GIS-O.

Chapter 3

Integration Strategies of GIS and Optimization Systems

3.1 Introduction
3.2 The importance of GIS-O integration strategies
3.3 The full GIS-O integration strategy
3.4 The loose GIS-O integration strategy
3.5 The tight GIS-O integration strategy
3.6 Comparison between the main GIS-O integration strategies
3.7 Potential applications of GIS-based optimization tools
3.8 Conclusion

3.1 Introduction

As outlined in the first chapter, GIS are getting pride of place in addressing strategic applications where data are structured as multi-layered themes in the database. GIS also offer numerous functionalities as collecting, analyzing and displaying data. GIS has proven its efficiency by solving different kinds of problems in different application domains which made it so popular and widespread. However, the future of GIS is full of challenges. One big challenge is the complex nature of some strategic problems that GIS must deal with and with which it often shows its limitations. In order to avoid such situations, the use of GIS must be consolidated.

An investigation of the existing literature reveals the need of integrating geomatics with other domains as statistics, marketing, decision making and transportation. The related integration frameworks are named, respectively: Geostatistics (Haining et al., 2010), Geomarketing (Faiz, 1999), Geodecisional (Faiz, 2005 and Zbidi et al., 2006), and Geotransportation (Kharrat et al., 2008). In fact, optimization seems to be a good partner for GIS helping it improve its performances and overcome its limitations. Optimization offers various techniques so that GIS becomes able to tackle and solve more complicated problems. We have shown through a study of some potential applications, the need for GIS. Such solutions are strongly dependent on the adopted techniques. To improve the quality of the generated solutions, we propose in this chapter a design of how to manage geographical data (GD) to be appropriately formulated using numerical approaches. Interestingly, GD are characterized by their large volume, dynamic aspect and multiple scales. Such features make the GD complex. Therefore, a first step shall be to specify the studied area and launch a query to highlight its required layers. Then, with reference to that data, an optimization problem is proposed. In the sequel, the study of the complexity and feasibility of the problem leads to a suitable choice of the related solution approach.

The existing GIS can show the potentials of an area under consideration: its proximity and neighbors, its shape and other related information that help the DM interact suitably with the environment to make the right decision.

This chapter takes place in the framework of integrating two systems belonging to different disciplines in order to strengthen the efficiency of the whole system. As we are addressing the integration of a GIS and an optimization tool, we call such system a "GIS-O" that can operate with respect to various protocols.

The overview of GIS-O covers three main classes of integration strategies namely the full, the loose and the tight integration strategies. We discuss the main advantages and drawbacks of each approach and dress a comparative study from both theoretical and practical points of view by exposing some potential applications. This survey makes easy to see that the more appropriate GIS-O integration strategy is application dependant. Special attention is dedicated to the class of transportation problems that will be the main potential application for which we propose a specific integration approach, designed as a decision support system, that manages the data from the geographical database and solves the obtained problem to be later plotted on the map.

This chapter is organized as follows. Section 2 shows the importance of integrating a GIS and an optimization solver in a common environment and highlights additional functionalities that offer such integration. We present in sections 3, 4 and 5 the main GIS-O integration strategies and their abilities and limits in taking in charge strategic and large sized tasks. Section 6 outlines a comparative study between these approaches based on previously underlined advantages and drawbacks. As researchers have paid a great deal of attention to that integration protocols, section 7 overviews some potential applications that retrieve the most pertinent integration strategies depending on the addressed problem.

3.2 The importance of GIS-O integration strategies

The GIS has proven its efficiency by solving a countless number of spatial decision problems, rising from various application domains. Thanks to these solving capabilities, the GIS is becoming increasingly popular and widespread. However, the future of the GIS is full of challenges. The most important and common challenge is related to data acquisition since it is often hard and expensive to collect (Cox and Gifford, 1997 and Faiz, 1999). Although data accessibility constitutes a continuous subject of debates among the GIS community, our research is not dedicated to question such issues nor propose solutions for it. An important and critical aspect consists in facing modern utilization of GIS by addressing real world problems. GIS is used to solve problems belonging to different domains, from which we state:

- Hydrology and water resource (Olivera and Maidement, 1999; Sui and Maggio, 1999 and Schumann et al., 2000).
- Environment and ecology (Groenigen et al., 1996; Marulli and Mallarach 2004 and Rossi et al., 2009).

- Waste management (MacDonald, 1996; Sumathi et al., 2007 and Alvarez et al., 2007).
- Urban planning (Gomes and Lins 2002; Wang et al., 2004 and Li, 2011).
- Transportation (Jankowski and Richard, 1994; Keenan, 1998; Jha et al., 2001; Huang and Pan, 2006 and Perpina et al., 2009).

These examples, among many others, clearly show how much GIS is useful and flexible in dealing with other domains. However, some limitations are to be underlined when handling some complex and hard analysis of tasks to provide the best decision (Parks, 1993, Albrecht, 1996; Bivand and Lucas, 1997 and Zhang and Grith, 1997).

Some sophisticated analysis necessitate advanced functionalities as:

- Generation of efficient solutions for loading and routing problems.
- Finding ways and channels of lower costs in complex networks.
- Capacity optimization of antennas in telecommunication networks.
- Fragmentation and packaging of data in networks.

A study of the literature reveals that, under such circumstances, using solely GIS does not provide satisfactory solutions. In fact, it becomes unable to meet the high expectations of the fast growing of GIS community.

For example, a complex landscape-ecological assessment has been performed in order to develop an efficient land-use planning of the region of Jeswitz situated in the northeast of Leipzig at Saxony in Germany. To this end, Graubaum and Meyer (1998) considered more specific objectives related to the assessment of soil erosion hazards by water, the water discharge regulation function, the groundwater regeneration function and the agricultural production function.

The authors pointed out that such complex task cannot be achieved using traditional maps generated by GIS, since they dispose of four goals to which variable weighting factors can be associated and this entails many potential results.

In order to face this challenge, efforts have been devoted to obtain satisfactory solutions. Since GIS is not able solely of performing complex spatial analysis, it seems very logical to think about integrating it with complementary tools.

Before thinking of what and how to integrate, it is important to understand first the meaning of integration itself: it refers to the process of bringing and linking together some components into a master system and making sure that they behave as a coordinate whole.

GIS have been integrated with numerous tools as optimization routines, statistical procedures, or visualization programs.

In this context, some authors (Muttiah et al., 1996; Grabaum and Meyer, 1998; Schumann et al., 2000 and Alvarez et al., 2007) proposed to combine optimization approaches with GIS. These related studies showed that the proposed integration constitutes a promising solution that makes easier and more efficient the resolution of complex real world problems.

Once the optimization system to be integrated with the GIS is chosen, the second concern is related to how to perform such integration. In general, three integration approaches are adopted to combine optimization tools with GIS namely, full, loose and tight integration strategies:

- **Full GIS-O integration strategy:** Optimization routines are developed within the GIS environment or conversely. In the latter case, GIS routines are embedded into the optimization tools.
- **Loose GIS-O integration strategy:** The GIS serves as a pre-processor and a post-processor to the optimization system.
- **Tight GIS-O integration strategy:** The integration of the optimization tools and GIS can be seen as a hybridizing of the two previous approaches.

There seems to be broad agreement that integrating GIS with optimization tools is quite an appropriate way to make GIS able to solve difficult and large sized problems in various domains. Still, achieving such integration between these two distinctive areas cannot be seen as an intuitive task or easy mission to accomplish. To do so, it should be defined and applied carefully in order to get accurate results. In fact, the importance of integration strategies can best be demonstrated by the growing number of scientific papers addressing the existing coupling approaches, their specificities, which approach to use, and in which situations.

Despite the differences in names and some details registered from one paper to another, three main integration strategies have been detected from surveying the literature and fortunately, have got the approval of the whole GIS community. These integration strategies are largely termed, as previously evoked, the full, the loose and the tight integration strategies (Goodchild et al., 1992; Stuart and Stocks, 1993; Fischer, 1994; Batty and Xie, 1994; Karimi and Houston, 1996; Bivand and Lucas, 1997; Brandmeyer and Karimi, 2000; Malczewski, 2006 and Vairavamoorthy et al., 2007).

We present, in what follows, these integration approaches by showing how they operate, their advantages and drawbacks. Then, we develop a comparative analysis based on some evaluation criteria in order to facilitate the process of choosing the right strategy in terms of the situation to be handled. This analysis is enclosed by stating some practical applications in various domains as hydrology and water resources, waste management, criminology and transportation.

3.3 The full GIS-O integration strategy

As its name might suggest, one system is fully integrated or embedded in the other. Applying this obvious definition to GIS and optimization software leads to the detection of two directions of full integration that can be achieved: either to embed the optimization routines into GIS software or embed GIS routines into the optimization software, as shown in figure 3.1. If we consider the first direction of the full integration (as reported in figure 3.1.(a)), implying that the GIS constitutes the main environment, DMs belonging to different domains can adopt such full integration as it offers many facilities in both ergonomic and interactive aspects. This can be an efficient way to overcome the complexity of some situations and satisfy the expectations of any DM regardless of his expertise.

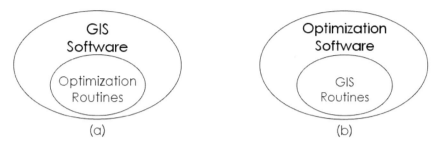

Figure 3.1 The full integration strategy.

Although we are interested in integrating optimization tools within GIS, many remarks might turn out to be valid for the second direction as the lack of ergonomy for the whole system and the complexity of the GIS.

3.3.1 Characteristics

Technically speaking, the full integration is referred to as a complete integration of optimization routines into GIS achieved using a programming environment.

In practice, this environment can be a GIS macro language as Avenue Macro Language (Zhang and Griffth, 1997) and/or advanced programming languages such as C/C++ or Java.

According to Huang and Jiang (2002), the GIS macro language is preferable since it ensures the stability and durability of the full integrated system, avoids problems of compatibility as well as the need for reprogramming all or some GIS functionalities.

Indeed, the full integration strategy provides a highly convenient platform to conduct the whole integration process from data selection and transformation stage to results display stage. Thus, using the embedded strategy, the DM can benefit a lot from the GIS visualization functions that result in various outputs such as chart, layout, 2D-maps and 3D-maps. Besides, for modern GIS, the visualization functionalities are no longer dedicated to mapping analysis but even better, they support spatial analysis (Dibiase, 1990 and Jiang, 1996). In the sequel, using GIS macro languages allows creating user-specified optimization routines that are meant to be integrated into the existing command set of the GIS software (Malczewski, 2006).

The full integration approach is also defined by Brandmeyer and Karimi (2000) as a master-slave relationship (the GIS being the master and the optimization component being the slave), in which case users recognize and interact only with the master while the slave provides valuable services without being aware of its existence.

In order to appreciate the use of the full integration strategy better, the following main advantages are printed out:

- **GIS strengths:** Since optimization routines are completely integrated into the GIS environment, they can benefit largely from the well-appreciated strengths of GIS as databases, visualization functions and analysis capabilities.
- **Easiness:** DMs are able to dress spatial analysis that perform optimization routines without ever leaving the GIS environment.
- **Rapidity:** The model is running faster and in a more efficient way because there is no data passing and all information is stored internally.
- **Accessibility:** Optimization routines are available for all users of the package.
- **Testability:** Full integration allows users to evaluate easily and efficiently different scenarios called *what-if* modeling by adjusting parameters and exploring the results of such changes immediately.

However, despite this undoubted advantages, some disadvantages of the full integration strategy must be outlined:

- **Component specificity:** Full integration results in a component specific property which means that the integration outcome strongly depends on types and versions of both GIS and optimization softwares. Thus, exchanging one tool by another or attempting to maintain the whole system is made very difficult.
- **Programming complexity:** To implement a full integrated model, a GIS macro language and/or advanced programming languages are required. This entails huge programming efforts

over a period of time to develop sophisticated algorithms for the integration.

- **Expertise:** The full integration approach necessitates an exhaustive understanding of each model before the integration as well as a complete knowledge of the programming language to be used.
- **Costly:** This approach might require major changes in the GIS software from vendor which will be very costly and not necessary for the majority of users (Goodchild et al., 1992).
- **Copyright:** To apply this strategy, the access to source code of both embedded and embedding models is required which is not, commercially speaking, always possible in case of proprietary systems (Fischer, 1994).
- **Maintenance:** In this topology, changing the master code in the GIS may result in changing the embedded model too. Such a task is often difficult.
- **Implementation complexity:** The use of a single computing system can be problematic whenever many optimization techniques are to be embedded in a GIS. This requires powerful and costly hardware (Arentze et al., 1996).

3.3.2 Potential applications

The advantages stated above have motivated some researchers to bring to life the full integration strategy by developing models based on its principles, while others prevent the possibility of disadvantages that can emerge from applying such a strong relationship between models to integrate. Despite this reticence against using full integration strategy, this latter turned out to be efficient in embedding simple mathematical models when convenient programming languages are available. However, it is to be noted that, in our context, very little effort has been done to fully integrate optimization software into GIS. In what follows, we enumerate some practical applications illustrating the use of the GIS-O full integration strategy:

- Rupp (1996) exploited the full integration approach by merging two mathematical dispersal models into the GIS to solve the problem of white spruce seed dispersal in the interior of Alaska. This model allows simulating the ecological system of seed dispersal upon a defined landscape unit and providing a valuable aid to the silvicultural decision making process.

- Gorokhovich and Janus (1996) proposed a full integration of an optimized phosphorus loading lumped-parameter into the GIS. The generated system provides improved performance analysis by supporting applications from watershed management.
- Huang and Jiang (2001) also developed a full integration of the topography-based hydrological model known as Topmodel within Arc View GIS. This integration allows a better exploitation of the GIS visualization and spatial analysis functions.

3.4 The loose GIS-O integration strategy

Unlike full integration strategy—which means completely integrating optimization routines into a GIS-, the loose coupling strategy aims to keep the two systems independent in the sense that no one embeds the other. Thus, the role of the integration process lies in the exchange of data files between the two systems as reported in figure 3.2, showing that output data from the first system are inputs for the second system (Malczewski, 2006).

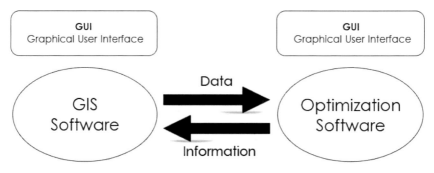

Figure 3.2 The loose coupling strategy.

3.4.1 Characteristics

Goodchild et al. (1992) viewed the loose coupling strategy as letting the GIS perform tasks for which it is best suited. When other services going beyond the capabilities of GIS are required, the optimization receives the GIS output, process the data and send back results to the

GIS. Consequently, the loose coupling approach takes full advantage of database and visualization tools of GIS as well as all possible optimization techniques that an optimization software can afford. Moreover, loosely-coupling GIS with any other external model (not only optimization tools) has been the foundation to building the GIS-based decision support systems.

Technically speaking, managing the exchange of data files between GIS and optimization tools is the primary issue in the loose coupling. Thus, enhancing the quality of this strategy relies on making the communication process faster, the conversion task more efficient and avoids problems of incompatibility. These goals have been the main concern of researchers while dealing with the loose coupling approach. The first related works used the ASCII files for data exchange. This approach was studied by Waugh (1986) for the design of the Geolink system and by Gatrell (1987) in coupling Odyssey and Glim softwares.

More recently, El Kadi et al. (1994) and Tait et al. (2004) proposed to use trigger functions, integrated into the GIS and the optimization component, to ensure an automatic transfer of data dynamically. The use of the loose coupling strategy is very widespread in the GIS community thanks to the following advantages:

- **Exchangeability:** Loose coupling strategy offers the possibility to exchange the GIS software or the optimization software with other tools and newer versions when needed or at will. Thus, the integrated system can be modified and updated easily without having to worry about the required time and programming complexity.
- **Independence:** Linking components by exchanging and converting data ensures the independence between systems and lets each one do what it is made and suitable for, without being influenced by any kind of restrictions that the other system might impose.
- **No component specificity:** Loose coupling results in a no component specific property which means that the integration outcome is independent of both the type of GIS and optimization softwares. Thus, such integration can operate with different GIS environments and any optimization software.

- **Flexibility:** Loose coupling is endowed by a considerable flexibility in dealing with GIS users that consists in offering the possibility to choose suitable tools to link with a GIS.
- **No integration time:** Loose coupling avoids redundant programming that permits time saving. Compared to other integration strategies, the time required for the loose coupling is insignificant.
- **Cost saving:** The two systems are linked with minor changes to their codes. Hence, the cost of the whole system is rather reasonable.

The remarkable advantages stated above must not hide the other inhibitor factors to use the loose coupling approach. We enumerate in what follows the main drawbacks facing the use of this strategy:

- **Unfriendly to user:** Loose coupling is known to be unfriendly to user in the sense that a DM have to interact with two different and separate interfaces and alternate between them which is a less comfortable situation.
- **Expertise:** Loosely coupling two systems often entails that the DM must be expert in both systems. Hence, the loose coupling requires from users much time and learning efforts leading to a resistance behavior to use the integrated system (Albrecht, 1996).
- **Complexity:** Data conversion between different existing formats can be tedious, error prone and loss of information is very likely to happen.
- **Low integration:** Loose coupling ensures only a partial integration due to the lack of available data conversion routines. This turns out to be less than the expectations of DMs.
- **No assistance:** The excess of flexibility left for the user can turn into a curse since it makes him completely responsible of his choices (models to integrate with GIS, tools, files formats, interface).
- **Network dependant:** The efficiency of this strategy relies on the quality and speed of the used network.

Before exposing some potential applications, it is worthwhile to note that the loose coupling strategy has a special case termed the GUI coupling: instead of dealing with two separate user-intefaces, a single GUI is implemented hiding the internal coupling method, as shown in figure 3.3. In addition to most of standard loose coupling advantages stated above, the GUI coupling strategy provides a user-friendly way for linking a GIS software and an optimization software. This is embodied in a single working environment where users are no longer aware of the locations and configurations of the systems being accessed. Thus, DMs are given the comfortable feeling of dealing with a single system (Berry et al., 1997).

Nevertheless, the easiness promised for users is often compensated by difficulties facing developers in implementing a single user interface for such disparate systems. This attempt results in increasing programming complexity and integration time.

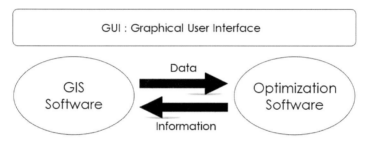

Figure 3.3 The GUI coupling.

3.4.2 Potential applications

In spite of the above deterring facts, the loose coupling strategy is the most adopted integration protocol of GIS and optimization as stated by many researchers, mainly those who have reviewed the literature about GIS integration issues (Bivand and Lucas, 1997 and Malczewski, 2006). In fact, such strategy avoids much programming and keeps the two systems as independent as possible, especially with the increasing number of potential models that can be linked to GIS. Besides, loose coupling makes the final product more profitable and corresponds to the current commercial trend. Some practical applications using this popular integration strategy are presented in what follows to develop a more complete view:

- Grabaum and Meyer (1998) linked a GIS disposing of assessment functionalities (as soil erosion hazards), groundwater regeneration and water discharge regulation, to a multicriteria optimizer. The obtained framework provides an efficient land use scenario for the north of Leipzig (Germany).

- GIS was used to identify agricultural and forest residue biomass locations. To do so, Perpina et al. (2009) solved a routing problem that minimizes concurrently distances, times and travel costs. The solution obtained via the Network Analyst solver consists in an optimal design of sites pathway for bioenergy plants.

- In hydrology and water resources, Schumann et al. (2000) developed a SDSS that detects the lumped catchment characteristics. They showed how loosely coupling an iterative procedure of numerical optimization with GIS can be beneficial to make such rainfall-runoff model more efficient.

- The simulated annealing metaheuristic is loosely coupled with GIS in order to find optimal locations for hazardous waste disposal in the pine watershed of India (Muttiah et al., 1996). The system offers to the policy maker the opportunity to select final sites taking into account social factors that cannot be handled by numerical models.

- A fuzzy multiobjective programming model was loosely coupled to a GIS in order to produce an efficient environment that decides which land use can be changed, maintains compatibility of land uses and controls physical conditions. To show the effectiveness of such integration, a SDSS that loosely integrates a GIS and a fuzzy multiobjective program was developed by Wang et al. (2004). They studied the design of an environmental planning for the specific case of the lake Erhai in China.

3.5 The tight GIS-O integration strategy

The tight strategy can be thought of as a hybridizing of the two previous approaches. In fact, tight coupling, also called close coupling strategy, expects from GIS and optimization tools not only sharing communication protocols but also user interface and common data storage (Batelaan et al., 1993). Such strategy avoids outcome incompatibilities. This means that under a tight coupling approach, the two systems must be linked within a common working environment (Sui and Maggio, 1999).

Therefore, as shown in figure 3.4, useful functionalities are embedded into the GIS using its macro language or more advanced programming languages (full integration part), while less demanded functionalities can still be accessed via the optimization software (loose coupling part).

The main strength of the tight coupling is the offering of a single GUI that interact either with the GIS part or the optimization part to handle complex tasks without paying attention to the belongingness of corresponding functionalities to either sub-system.

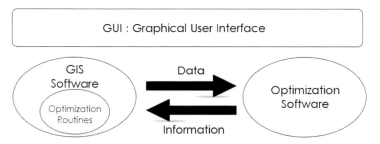

Figure 3.4 The tight coupling strategy.

3.5.1 Characteristics

From the technical side, tight coupling requires the development of models with dynamic link libraries (DLLs) using programming languages such as C/C++ or Fortran (Ding and Fotheringham, 1992 and Batty and Xie, 1994). To sum up, according to Pullar and Springer (2000), major features of tightly coupled optimization tools with GIS are: (i) implementing a common GUI, (ii) including an optimization toolbox of basic functions, often needed by users, into the GIS, and (iii) developing a shared library to exchange data between the GIS and optimization tools. Compared to loose and full integration strategies, the use of the tight coupling approach is encouraged by the advantages fusion of the two previous approaches:

- **High level of integration:** Compared to the full integration, the tight is a more reasonable and meaningful approach that avoids embedding integration solution by avoiding fully integrating all required optimization techniques into the GIS. As it is a more realistic option, it allows integrating the basic and common

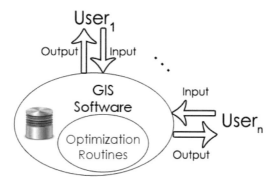

Figure 3.5 Architecture of the centralized approach.

tools frequently needed in order to reduce the complexity of the output system (Tait et al., 2000).

- **Low data transfer complexity:** Compared to the loose coupling approach, the effort of editing and converting data files is considerably reduced since a number of common optimization tools are embedded into the GIS.

- **Reusability:** Tight coupling is designed as a generic framework that allows the reusability for several related applications necessitating minor and simple adjustments.

- **Rapidity:** Since most of the used optimization functionalities are hosted in the GIS, the transfer time required to accomplish optimization operations is reduced.

Being a hybridization of loose and full integration strategies helps the tight coupling approach to remedy to some of their critical disadvantages. However, this fact does not spare this approach from being criticized. Major inherent drawbacks are reported below:

- **Component specificity:** As for the full integration strategy, the tight coupling approach inherits the component specificity drawback due to the close linkage in the GIS system.

- **Programming complexity:** This strategy requires the GIS macro language and rich programming languages to support high integration protocols. Available programming tools are still incapable of handling complex scientific modeling and iterative processing tasks.

- **Storing difficulty:** Storing intermediate results rouses a problematic issue in applying the tight coupling approach because GIS languages encounter problems in dealing with temporary variables (Batelaan et al., 1993).

- **Communication complexity:** Lacks can be depicted in accomplishing some tasks due to the complicated communication process between GIS macro languages and developed DLLs.
- **Data constraints:** Formatting requirements for the transfer are to be taken into account for the linkage of heterogeneous environments which affect computational performances.
- **Need for spatial SQL:** The need for developing a standard query language for spatial data that enables all users to access this data without any need to know the specific structure used in the GIS database (Goodchild et al., 1992).

3.5.2 Potential applications

In practice, tight coupling is seen by many authors (Ding and Fotheringham, 1992; Batty and Xie, 1994 and Pullar and Springer, 2000) as an efficient integration strategy that had been widely used for analysis and modeling purposes in GIS environments.

Some of the uses of tight coupling are illustrated by the following:

- Vairavamoorthy et al. (2006) developed a software that couples a GIS with three optimization models tightly using Avenue macro language and DLLs. This software is designed to assess the risk of contaminant intrusion into water distribution systems from pollution sources such as sewers, drains and ditches. Indeed, it provides to DMs a reliable basis for assigning investment priorities related to water quality management.
- A GIS prototype to simulate land conversion was developed by coupling Arc/Info GIS tightly with a multicriteria modeling for cellular automata. Being simple and convivial, the application provides urban planners with more realistic solutions consisting of transition rules for cellular automata (Wu, 1998).
- In traffic incidence management, Huang and Pan (2006) developed a response tool based on a tight integration of a GIS with LINDO optimizer and a traffic simulator. Coupling these different models Tightly has considerably reduced the response time and provided an optimal dispatching of response units.

3.6 Comparison between integration strategies

In the foregoing sections, a deep analysis was driven to evaluate the different integration strategies. On the one hand, each of them offers numerous advantages that encourage its performing. On the other hand, bothering disadvantages were detected urging the reconsideration of one choice or another.

In practice, understanding characteristics, advantages and drawbacks of each strategy is never enough to make the right decision. It is necessary to admit that selecting the appropriate integration approach is largely correlated with many factors from which we state:

- DMs priorities and potential trade offs.
- The objectives to be reached.
- The complexity of models to be integrated.
- The application domain for which the integration takes place.

Investigating the literature reveals that very few papers have justified the use of an integration strategy and even less researches have addressed how to decide given a specific situation and goals to be achieved.

Being aware of this fact, we propose in what follows to assist the DM in his primary and critical mission that entails the quality and efficiency of the integrated system.

To this end, we propose first a comparative analysis based on a number of technical criteria deduced in table 3.1. Then, we outline, in table 3.2, a comparison based on a centralized/decentralized architectures adopted to implement the integrated system.

3.6.1 Comparison-based technical criteria

We detail, in what follows, the most relevant properties printed out in table 3.1 for comparing GIS-O integration strategies:

- **The level of integration:** It measures the degree of cohesion between the modules of the integrated outcome.
- **The speed of data transfer:** That evaluates the rapidity of data transfer between the modules of the integrated system.

Table 3.1 Comparison of technical criteria for integration protocols.

Property	Full	Loose	Tight
Level of integration	Complete	Low	High
Speed of data transfer	Fast	Low	Moderate
Component specificity	Component specific	No component specific	Component specific
Programming language	GIS macro and/or advanced languages	Advanced languages	GIS macro and advanced languages
GUI	Enriched	Not enriched	New/Enriched
Integration cost	High	Low	High
Data storage	Common	Independent	Common /Independent
Integration time	Considerable	Short	Moderate
File conversion	Internal GIS formats	GIS import/ export formats	Internal GIS formats and GIS import/ export formats
User expertise	GIS	GIS and optimization	GIS

- **The component specificity:** Expressing the dependency of the output system on a specific GIS.
- **The programming language:** It refers to the languages used to implement the integration process (GIS macro languages and/ or advanced languages).
- **The GUI:** Which informs about the design of the used graphical user interface that hides the integrated systems. This can be done by either enriching the existing interface (the full integration strategy) or by providing a specific GUI that manages the whole environment (loose and tight integration strategies).

- **The integration cost:** It assigns a cost level for operationalizing each integration strategy.
- **The data storage:** That describes how geographical and optimization data are stored in separate or common storage supports.
- **The integration time:** It is a qualitative measure of the time response for implementing each integration strategy.
- **The file conversion:** Describing how data files can be converted between the modules of the integrated system.
- **The user expertise:** It indicates how much expert should the DM be. This level of expertise informs about the efficient use of the integrated system.

3.6.2 Comparison-based centralized/decentralized architectures

A centralized architecture is based on a central decision making process. In computer terms, it means that the software application is hosted in a central server. This entails that a unique and central database is available for all users of the whole system. Key advantages of this architecture are manageability and coherence. However, when the organization is composed of many departments, the central control becomes inefficient.

Meanwhile, a decentralized architecture allows dispersing the decision making process. The system adopting this architecture is divided into sub-systems that work independently while communicating via a connected network. In addition, each sub-system manages its own database specialized in its activity. It is useless to mention that these specialized databases are connected to a central database which manages the data exchange between the different sub-systems.

Compared to the previous approach, the decentralized architecture guarantees simplicity, performance, scalability and reliability.

In the specific context of deciding which integration strategy to choose, it is of primary importance to select the suitable architecture to implement for the integrated system:

- **The full integration strategy:** It entails embedding optimization routines into the GIS. Thus, the integrated system is the GIS itself enriched by optimization functionalities and only one computer system is required.

  Consequently, the full integration strategy is highly recommended for a centralized architecture that imposes a centralized decision making process. Moreover, the centralized approach is often applied when a project is to be developed for a specific domain.

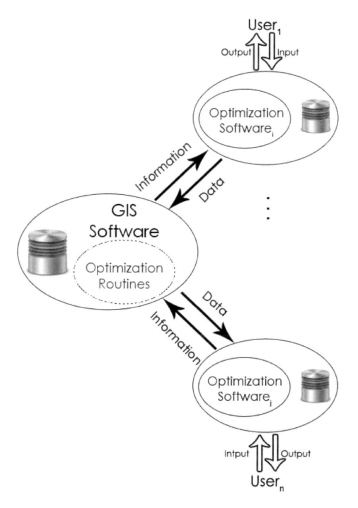

Figure 3.6 Architecture of the decentralized approach.

Thus, a full integration strategy is appropriate since there is no need to worry about the reusability of the application.

- **The loose integration strategy:** As GIS and optimization softwares operate independently, the data exchange is ensured by file transfers. This organizational mode corresponds to the decentralized architecture.

Another strong argument to design the loose integration according to the decentralized architecture is the structure of the whole framework as it encompasses numerous separate sub-systems requiring different optimization tools. In such a case, each sub-system is constraintless to use the set of optimization tools that respond to its specific needs.

- **The tight integration strategy:** It takes into consideration two types of optimization modules: (i) standard modules consisting of numerous common functions easily integrated in the GIS, (ii) specific modules involving a personalized set of functions. In this case, a combination of a centralized and decentralized architectures is required.

Table 3.2 Comparison-based centralized/decentralized architectures.

	Centralized	Decentralized
Full	Highly recommended	Not recommended
Loose	Not recommended	Highly recommended
Tight	Recommended	Highly Recommended

3.7 Potential applications of GIS-based optimization tools

We have depicted in the foregoing sections the motivation behind combining a GIS and optimization techniques and surveyed the most important integration strategies that allow such combination. However, performing the integration for its own sake is rather meaningless. In fact, integrating GIS and optimization tools becomes meaningful and extremely beneficial when applied to solve real world applications. As it is the case, various potential applications can perform better while handled by an integration framework. In this section, we discuss the application of GIS-based optimization tools in numerous fields of study as hydrology and water resource, waste management, criminology and transportation. Table 3.3 summarizes, for each application, the use of GIS, the applied optimization tools and which integration strategies can be beneficial.

3.7.1 Hydrology and water resources

Water is one of the most relevant natural resources for which many scientists brought a great deal of attention from many viewpoints. From a statistical standpoint, water is evaluated spatially and temporally along its hydrological cycle.

This spatio-temporal characterization is an incentive to perform a GIS-based application that handles such structured data easily. Until 1980, GIS and hydrology were viewed as separate disciplines that evolve without any interaction. Henceforth, both hydrologists and GIS users became aware of the mutual benefits they could gain from integrating the two disciplines (Sui and Maggio, 1999).

Chow et al. (1988) proposed a classification of hydrological models in terms of the randomness, the space and the time. The combination of these parameters have led to define eight hydrological models that can apply GIS. However, traditional GIS are unable to handle hydrological modeling and to deal with the complexity generated by each model. That's why, the GIS should be consolidated by optimization tools in order to provide hydrological modeling with best knowledge of the terrain. Conversely, hydrological modeling allows GIS-based optimization to increase its analytical capabilities and expand its utilization (Goodchild et al., 1992; Fotheringham and Rogerson, 1994 and Singh and Fiorentino, 1996).

The following examples illustrate the application of GIS-based optimization systems in the hydrological context:

- A software was developed by Tait et al. (2004) to identify the optimum locations for new boreholes in order to exploit groundwater resources efficiently and predict potential contaminant concentrations in urban areas. This application, called Borehole Optimization System (BOS), is composed of three optimization models loosely coupled with the GIS.

- Schumann et al. (2000) also applied a loose coupling approach to link the GIS and optimization tools (numerical optimization) with three semi-distributed hydrological models in order to build a conceptual rainfall-runoff modeling. This framework offers a better estimation of the input parameters for different catchments, given a specific region. Using these information, an exhaustive description of the spatial catchment heterogeneity is obtained.

- A full integration of the epsilon constraint modeling in the GIS was developed by (Maeda et al., 2010) to assess water quality of rivers under hydro-environmental uncertainty and control waste loads from source pollutants. This model was used for the Yasu river in Japan. It turned out to be a valuable tool for improving water quality and optimizing the total load of the river.

Technically speaking, the integration of GIS and optimization tools in the hydrological domain was structured with regard to all integration strategies.

3.7.2 Waste management

Waste designates any unwanted or useless materials that one must get rid of. The modern society produces many waste types, namely, municipal solid waste as household waste, bio-medical waste including clinical waste, hazardous waste as industrial waste and special hazardous waste as radio-active and explosive waste.

In the past few decades, the urbanization phenomenon has fastened in such a way that huge amounts and various types of waste are being generated every day. As is the case, the problem can be reduced to a task sequencing of collection, transport and disposal of waste. The consequences of such problem are environmental pollution and related diseases. Therefore it is required to develop a sufficient and sustainable waste management system (Sumathi et al., 2007). As such, applying a GIS-based optimization framework to handle waste management is very promising (Muttiah et al., 1996; Ghose et al., 2005 and Alvarez et al., 2007). Following are some applications related to waste management:

- GIS was used to optimize siting decisions of municipal solid waste landfill taking into consideration important criteria such as water supply resources, land use, sensitive sites and air quality. The integration of GIS and optimization tools was achieved through adopting loosely the coupling approach (Muttiah et al., 1996 and Sumathi et al., 2007).
- A GIS routing model was proposed to optimize the collection and transport of solid waste to the landfill using the loose coupling strategy for the integration issue. This constitutes an important component of sustainable waste management as it offers to waste management planners and authorities considerable gains in time and money (Ghose et al., 2005).
- A GIS-based optimization system was developed by Alvarez et al. (2007) to calculate the number of containers needed in the streets of a specific area, their suitable locations and a design of pathways for collecting waste while minimizing the time required for delivering goods to all destinations. The proposed system was designed as a loose coupling of the GIS with the minimal generator tree optimization algorithm.

As the above examples might reveal, most integration of GIS and optimization tools in the waste management field is based on the loose coupling approach.

3.7.3 Criminology

Criminology is an interdisciplinary field specialized in studying causes and motives of criminal behavior in the hope of controlling and reducing crimes. To achieve such an important and security-related objective, criminologists are trying to perform their analysis based on many techniques such as measures of spatial autocorrelation, kernel density smoothing and hierarchical cluster analysis (Craglia et al., 2000; Harries, 1999 and Grubesic, 2006). In addition to these statistical tools and given the fact that integrating place and space constitutes the new trend in crime analysis, criminologists could not ignore the growing necessity of using GIS enriched with optimization solvers. In fact, GIS-based optimization handling crime factors has interested crime theorists and practitioners as police administrators, politicians and regional planners (Harries, 1999). Hereby are some illustrations of GIS-based uses in the criminology domain.

- With its cartographic interpretation and crime mapping analysis, the loose integration is performed to provide estimations for time and place for a crime before it happens (Bowers et al., 2004 and Grubesic, 2006).
- A tight combination of a GIS and an integer programming solver is designed to determine an optimal spatial distribution of police patrol areas by maximizing the backup covering location (Curtin et al., 2007).
- A loose integration approach was developed using aggregate crime datasets on which GIS and trace techniques, as the optimization part, were applied in order to represent the spatial distribution of these datasets. This approach offers a deep analysis of the complex crime behavior (Philips and Lee, 2010).
- GIS combined with heuristic methods was frequently used in law enforcement aiming to minimize the total traveled distance to service calls (Mitchell, 1972).

The integration of GIS and optimization tools to improve crime analysis is mostly designed as a loose integration approach.

3.7.4 Transportation

Transportation is viewed as a strategic decision problem that employs mainly resources and a spatial infrastructure (Kharrat et al., 2008). Transportation of items generally requires vehicles, as resources. The spatial infrastructure is related to roads, customers and depot(s)' locations.

The related literature reveals that traders, retailers and DMs in the supply chain brought a great interest for transportation problems. As a result DMs, analysts and planners have thought of ways to provide reliable transportation services and concurrently minimizing the overall traveling cost (Arampatzis et al., 2002). As a transportation problem requires mainly spatial data, it is obvious a

GIS should be used as it consists of a powerful tool for representing a transportation network more realistically. In addition, it can improve the integrity of transport database, ameliorate user-interfaces and offer a cartographic display of model results. Thus, transport modeling combined with GIS can enhance its role as a promising decision support system.

Although GIS is considered as a powerful tool, large-scaled transportation problems that require more sophisticated spatial analysis and computation algorithms not supported by GIS. Therefore, the GIS needs to increase and optimize its analysis performance by integrating optimization tools.

A literature survey by Malczewski (2006) from 1990 to 2004 showed that, among more than 13 application domains of GIS, transportation holds the second highest percentage after the environmental domain. Such a deduction is an incentive to and establishes the GIS-based optimization system is a great partner for transportation helping it solve complex variations. Among transportation variations, the vehicle routing problem (VRP) is one of the most studied class of problems due to its relevance in modeling and solving real situations. VRP refers to the supply of a number of customers at known locations with their specific demands using a fleet of vehicles subject to some constraints such as vehicles capacities, route lengths and delivery time (Tarantilis and Kiranoudis, 2002). A literature review due to Laporte and Osman (1995) identifies 500 articles more or less contributing to the routing field. From the numerous VRP variations, we can cite the open vehicle routing problem, the capacitated vehicle routing problem and the vehicle routing problem with time windows. Practically speaking, some uses of GIS-based optimization in transportation modeling are illustrated by the following applications:

- A GIS-based optimization system, designed as a SDSS, is used to route and schedule vehicles. Since routing problems have generally a high computational complexity, various metaheuristics are applied using, in most cases, the loose coupling strategy (Keenan, 1998; Tarantilis and Kiranoudis, 2002 and Tarantilis et al., 2001).

- Huang and Pan (2006) proposed a tight integration of three components consisting in a GIS, a traffic simulator and the LINDO optimization solver as a framework for managing traffic incidents in urban areas that improves time and solution quality.
- Sutapa and Jha (2011) developed a loose integration of a GIS and a genetic algorithm metaheuristic for rail transit alignment in which the optimal station locations are to be determined.
- Another application of GIS-based optimization in logistics and transport strategies tries to locate an optimal network of bioenergy plants within a specified region followed by a detailed study of the corresponding transportation network in terms of time, distance and transport costs (Perpina et al., 2009).

The integration of GIS and optimization in the transportation modeling is achieved according to all integration strategies. The loose coupling approach is used, frequently followed by the tight, then the full.

3.7.5 Summary of GIS-O applications

We report in table 3.3 each of the above GIS-O applications in terms of the following items:

- **GIS:** The main tasks to be accomplished by the GIS.
- **Optimization tools:** The adopted techniques to solve the corresponding problem. Depending on the problem difficulty, either exact or heuristic approaches are to be developed.
- **Integration process:** GIS-O integration protocol(s) are enumerated depending on the addressed problem.

3.8 Conclusion

We investigated in the present chapter the integration of two strategic environments, GIS and optimization softwares, that can help solving a countless number of real applications. An exploration of the existing literature reveals that three main integration strategies are of interest: the full integration, the loose coupling strategy and the tight coupling strategy. A comparative analysis between these integration protocols has shown that there is no best strategy, but it is rather case

Table 3.3 Summary of GIS-O-based potential applications.

Application	Hydrology and water resources
GIS	• Develop conceptual models as rainfall-runoff
	• Model a contamination-risk assessment
Optimization tools	Threshold based algorithms as the epsilon-constraint approach
Integration process	Full, loose and tight
Application	Waste management
GIS	• Optimize the collection and transport of waste
	• Identify the optimal number of containers and their locations
Optimization tools	Minimal generator tree and simulated annealing
Integration process	Loose
Application	Criminology
GIS	• Detect crime hot spots in areas registering high level of crimes
	• Predict when, where and why some crimes are likely to happen.
Optimization tools	Maximal covering models
Integration process	Loose
Application	Transportation
GIS	• Solve vehicle routing and scheduling problems
	• Pick up and delivery systems
Optimization tools	Minimum cost path algorithms and genetic algorithms
Integration process	Loose and tight

dependant. The choice of the suitable integration depends on the difficulty of the problem, the centralized or decentralized architecture to be adopted, the need of prompt responses and the nature of the data to be handled. One aspect that we would like to explore in the next chapter is the study of the VRP where data are performed from a GIS. This study, shows the effectiveness of integrating these two environments in the resolution of such complex class of problems.

A GIS-O Framework for the Vector Loading Distance Capacitated Vehicle Routing Problem

4.1 Introduction

In the field of transportation, the design of the delivery process, if well scheduled, is beneficial in various contexts as supply chain management and logistics. The delivery process consists in distributing orders from suppliers to some geographically dispersed demand points. The main concern in a distribution network is to specify the most cost-effective itinerary for delivering goods and commodities from a set of warehouses to customers. Such a process can be viewed as a vehicle routing problem (VRP), one of the widely studied combinatorial optimization problems, due to its relevance in modeling real-world applications.

The VRP requires further information as available routes and the traffic in the addressed zone. As such, the integration of this class of problems in a GIS according to the full, the loose or the tight strategy makes its modeling more realistic by providing more reliable solutions. In fact, many studies in the scope of integrating VRPs and GIS were proposed (Ruiz et al., 2004; Mendoza et al., 2009 and Santos et al., 2011). Alternative variations of the VRP were also investigated as the capacitated location-routing problem (Lopes et al., 2008) that consists in a combination of the facility location problem and the VRP. Spatial decision support systems (SDSSs) were developed to solve such routing problems and to handle such integration (Li et al., 2004 and Ruiz et al., 2004).

Another issue of great relevance in industrial firms concerns the loading problem (LP) that stows a set of items into bins while trying, in most cases, to minimize the number of bins. Interestingly, a high employment of the applied transportation capacities ensures significant cost savings. This class of problem is known in the literature to be *NP*-hard, and this makes computer packing heuristics perform consistently well.

Within these two problems, it is noticeable that the main aim of the DM is the minimization of his total cost. Generally, VRP and LP are evoked apart. However, in most practical situations, firms need to handle both problems giving rise to a master problem that provides a more balanced routing solution and prevents the loading of vehicles.

Thus, a new class of routing problems that we call the vector loading distance capacitated vehicle routing problem (VL-DCVRP), arises. The VL-DCVRP consists in combining the VRP with capacity and distance constraints in the one hand, and the LP for the vehicles, in the other. The main objective of the problem is to minimize the total cost while considering a set of structural constraints.

Such a problem falls in the GIS-O framework as it requires both geographical and optimization environments. The related solution approach benefits mainly from GIS functionalities and suitable optimization tools yielding to a fast convergence to satisfactory solutions.

We address in this chapter the VL-DCVRP, applied in real-world applications to manage efficiently a fleet of vehicles in an industrial company by their loading in terms of their routing.

We propose to resort to the GIS environment for the extraction of suitable information as an input to optimization routines for solving the VL-DCVRP. To do so, the optimization part operates, depending on the problem-size, by generating either an optimal or a near optimal solution using an exact algorithm or an approximate approach. The obtained solution is then cartographically visualized by the GIS that assigns a map for each vehicle, showing its pathway.

This chapter consists of the following sections. In section 2, the fleet management is defined as the general context of transportation followed by the geotracking technique, used mainly for dynamic routing problems. In section 3, an overview of the main VRP variants is discussed and a classification, depending on the involved set of constraints, is stated. Section 4 details the main features that characterize the VL-DCVRP, then its mathematical formulation. Illustrations by numerical examples are also interpreted. The loose integration strategy for the VL-DCVRP is outlined in section 5. To operationalize the GIS-O loose integration, we apply QGIS as the GIS part and the alternative use of the CPLEX solver and the tabu search metaheuristic, as the optimization part. Illustrations on a real map are also proposed to show the effectiveness of the GIS-O loose integration strategy.

4.2 General context

As GIS are becoming suitable frameworks to manage strategic routing tasks and to make the delivery process more efficient, it is necessary to define some related concepts. A special appeal to the fleet management, which is the main subject of the present chapter, was noted in the GIS environment.

For routing problems, specific techniques are to be performed as the geotracking which provides accurate information about coordinates of vehicles.

These two concepts help the DM handling all concerns related to vehicles as the track of information about the state of vehicles and the determination of optimal pathways that can be well modeled by the VRP or one of its variants.

4.2.1 The fleet management

A particular attention has been given to the utilization of GIS in various disciplines as agriculture, forestry, climatology, robotics, telecommunication, civil engineering, computer sciences and logistics. In fact, the GIS offers multiple functionalities showing its ability to handle spatial and spatio-temporal data, as the management of resources (especially vehicles) for the control of the stowing and the design of the routing. In this context, a GIS-based application for controlling a fleet of engines as trucks, crushes and stockpiles was developed by Beaulieu and Gamache (2006). Peters et al. (1996) and Awasthi et al. (2011) proposed, for the industrial framework, an interactive system for managing a fleet of automated guided vehicles that provides satisfactory solutions.

Fleet management can also be applied in dynamic environments where vehicles and their related data vary along a time interval. Customers demands generally vary in dynamic frameworks depending on the state and the time period. This version was introduced by Dantzig and Fulkerson (1954), then by Topaloglu (2006).

Fleet management can cover many perspectives for the vehicles (tracking, maintenance, financing and controlling). We can refer to Peters et al. (1996) and Awasthi et al. (2011) for a large variety of fleet management GIS-based applications.

4.2.2 The geotracking

One of the powerful techniques in fleet management, used mainly in dynamic environments, is the geotracking. This technology provides information during the routing of vehicles within a time window. Figure 4.1 illustrates the data exchange between the treatment center and vehicles equipped by terminals such as a global positioning systems (GPS).

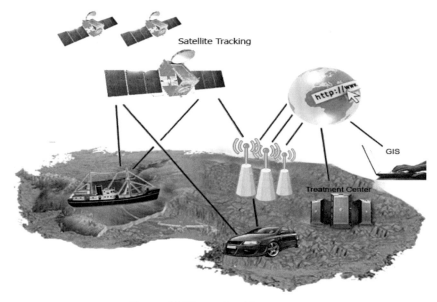

Figure 4.1 The geotracking process.

Color image of this figure appears in the color plate section at the end of the book.

The geotracking contributes in the minimization of the delivery cost as it permits:

- Receiving advanced information related to freight and security status.
- Controlling clearly the location and the quality of goods in a real-time.
- Recovering stolen vehicles by a fleet security.
- Gathering all vehicle's information such as the fuel consumption and the mileage.
- Displaying the itinerary to be adopted.

Meanwhile, some limitations of the geotracking are to be overcome such as:

- Generating the optimal itinerary.
- Finding strategies to avoid traffics.
- Designing a profitable loading process that depends on vehicles capabilities.

Generally, the geotracking technique is used when real-time information are to be extracted for managing the routing of vehicles known as the VRP. In this context, the GIS should be enriched by optimization functionalities in order to provide shortest paths.

4.3 VRP variants

Fleet management covers many topics for handling vehicles, specifically the delivery of goods which can be suitably modeled as a VRP.

The VRP and the LP were investigated separately in the literature. More recently, several studies looked into combining them, and numerous variations have been considered. Vigo (1996) studied the VRP with a loading constraint, known as the capacitated VRP (CVRP), that tries to satisfy a set of demands already ordered by geographically scattered customers using a set of vehicles based on a central depot. The main objective of the CVRP is, generally, to minimize the travel cost (Ai and Kachitvichyanukul, 2009), or alternatively to minimize the travel distance. Introducing distance constraints in the CVRP yields to the DCVRP which is about the delivery of cargo to customers while respecting distance thresholds for each vehicle.

Gendreau et al. (2006) introduced the two-dimensional CVRP (2L-CVRP) in which items are evaluated by their lengths and widths in addition to their weights. It consists in loading a set of geometric items into vehicles, then delivering them to customers. The objective is mainly the minimization of the travel cost expressed in terms of used vehicles (Zachariadis et al., 2009 and Duhamel et al., 2011).

If we speak about two independent evaluations of the set of items as the volume and the weight, the problem corresponds to a vector loading. Indeed, the consideration of distance constraints brings to life a new version of the VRP that we call the vector loading distance capacitated VRP (VL-DCVRP). Numerous variants for the VRP can be printed out:

- **The CVRP:** This version of the VRP was studied by Vigo (1996), then by Laporte et al. (2000). It is about fulfilling the known demands of geographically scattered customers by a set of vehicles based on a central depot. The main objective is to minimize the travel cost while respecting the routing and capacity constraints.

- **The DCVRP:** Once distance constraints are added to the CVRP, the problem becomes a DCVRP. It is about the delivery of cargo to customers while respecting distance thresholds for each vehicle.
- **The VL-CVRP:** We speak about vector loading when two evaluations are taken into account in the description of items. If the evaluation corresponds to the length and the width, the problem is said to be 2L-CVRP. This version of the problem was introduced by Gendreau et al. (2008), then studied by Zachariadis et al. (2009) and Duhamel et al. (2011).
  The problem consists in loading a set of bi-evaluated items into vehicles, then delivering the cargo to customers.
  The minimization of the travel cost is expressed alternatively by the minimization of the number of used vehicles.
- **The VL-DCVRP:** This version includes the whole set of constraints. It handles a bi-evaluation of items to be loaded into capacitated vehicles and delivered to geographically dispersed customers within a specific area (Iori et al., 2007). The main objective is to minimize the travel cost expressed in terms of the number of used vehicles.

Regarding the above variations of the VRP as clearly shown in figure 4.2, it is noticeable that the VL-DCVRP encompasses the entire set of constraints that are classified as:

- Circuit requirements
- Visiting each customer once
- Path continuity
- Subtour elimination
- Volume constraints
- Weight constraints
- Distance constraints

We propose, in the next section, to state a detailed description of the VL-DCVRP followed by its mathematical formulation. A numerical illustration of the mathematical model is presented and solved using the CPLEX solver.

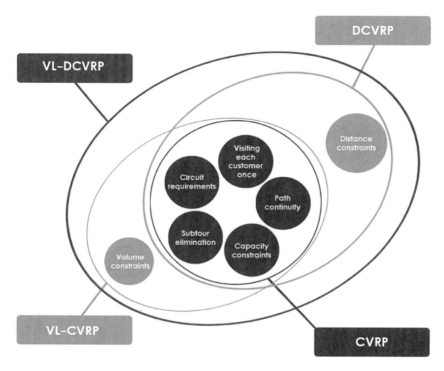

Figure 4.2 VRP variations.

Color image of this figure appears in the color plate section at the end of the book.

4.4 The VL-DCVRP

Due to its relevance in modeling in a more realistic way the loading and delivery of items, we propose in this section to state a detailed description of the VL-DCVRP, then to develop its mathematical formulation. The model is a single objective linear programming optimization problem that contains seven sets of constraints, as described earlier. As the number of constraints related to the subtour elimination is exponential, the problem becomes hard to solve by means of an exact method (as the branch and bound algorithm) especially for large sized instances. We detail the main steps of a numerical illustration and how long will smallsized VL-DCVRPs take to be solved.

As an incentive toward a GIS-O integration, we underline some limitations when generating numerical solutions that are far from any practical tendency as they lacks of details that can help the driver and the scheduler considerably in setting a detailed plan of the routing process. In fact, a cartographic output that clarifies and operationalizes the loading and mainly the efficient delivery of items is greatly beneficial.

4.4.1 Problem statement

The VL-DCVRP is a fleet management problem described in figure 4.3 as a two-step process:

1. **The loading step** that consists in stowing the cargo into vehicles. Such scenario is modeled as a LP.
2. **The routing step** that determines optimized routings of the vehicles along the road network while trying to satisfy customers' requirements. This stage corresponds to the VRP.

Initially, the LP and the VRP were studied separately. Then, as the literature reveals their closeness and complicity in specifying many practical situations, a combined version of the two sub-problems represents the right formulation that well reflects such situation (Iori et al., 2007).

The main advantage of the VL-DCVRP is that, by tackling the information on the freight to be loaded, one can construct more appropriate routes for the vehicles with cost savings. But in return, there is a rise in the combinatorial difficulty of the problem.

The VL-DCVRP involves a set of objects already stored in a depot to be loaded into vehicles, then delivered to some geographically dispersed selling points, in order to fulfill known customer requests. The scenario of VL-DCVRP, as described in figure 4.3, is stated as follows:

- The choice of suitable vehicles ensuring some resource limitations as the total distance and the quantity of the freight.
- The cargo loading of orders in vehicles from a single depot.
- Routing schemes for the delivery to customers correspond to orders already packed in vehicles.

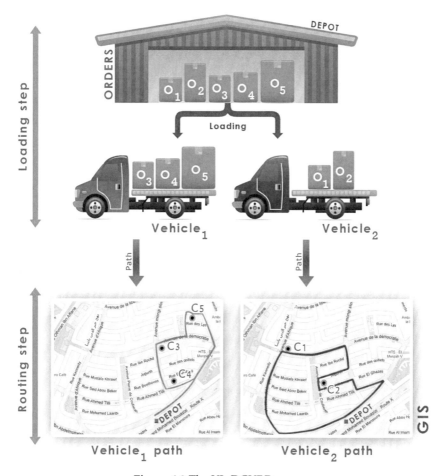

Figure 4.3 The VL-DCVRP process.

Color image of this figure appears in the color plate section at the end of the book.

Thereby, the problem is about planning the loading of each vehicle, then its round in such a way that it minimizes the total cost while considering some rules:

- Every order has to be carried by exactly one vehicle.
- The weight capacity of each vehicle should be respected.
- The total volume of orders packed in a vehicle should be in the volume interval of that vehicle.

- Each vehicle should not exceed a prefixed distance and cannot accept a path less than a minimum threshold.

The whole scenario is summarized in figure 4.3 where the inputs are information related to the set of items and vehicles configurations. The outputs of the optimization process are the assignment of items to vehicles and the design of their pathways.

We propose to solve the VL-DCVRP using a GIS in such a way to track the related inputs from the GDB. Then, an appropriate optimization approach has to be applied to get the optimal solution that will be later highlighted in a cartographic format.

4.4.2 Mathematical model of the VL-DCVRP

As the VL-DCVRP is a two-compound problem that consists in loading vehicles and proposing pathways with a minimal cost incurred by the used vehicles, it is important to pinpoint the main components that characterize a right formulation of the problem. The remainder of this section shows how to model the VL-DCVRP using the graph theory background that helps to outline its main components and specify the inputs and outputs of the mathematical model.

The VL-DCVRP can be defined as an undirect connected graph $G = (V,E)$ where:

- **The set of nodes:** $V = \{0,...,m\}$ in which 0 designates the depot D and indices $i = 1,...,m$ refer to customers. Thus, $|V| = m + 1$.
- **The set of edges:** $E = \{\{i,j\} \mid i,j \in V\}$ that expresses the possibility of direct routes between each pair of customers, or between the depot and a customer. Each edge $\{i,j\}$ is weighted by a distance d_{ij}. If the graph is complete the number of edges is:

$$|E| = \frac{m \times (m + 1)}{2} \tag{4.1}$$

Figure 4.4 corresponds to a complete graph with one depot and 4 customers, in which case 5 nodes are labeled from 0 to 4. Each edge $\{i,j\}$ is characterized by its distance d_{ij}. Hence,

- $V = \{0, \dots, 4\} \implies |V| = 5$ vertices.

- $E = \{\{i, j\}| \ \forall i, j \in V\} \implies |E| = \frac{4 \times 5}{2} = 10$ edges.

As the objective function expresses the minimization of the total cost, a unit cost c^k per Kilometer, depending on the used vehicle, is applied to compute a cost matrix M^k_{costs}, based on the distance matrix that contains all values of d_{ij} reported from the graph G. The distance matrix $M_{distances}$ (described in figure 4.5) is a symmetric square matrix of order m + 1 since $d_{ij} = d_{ji}$. Furthermore, all elements in the diagonal are set to 0, that is $d_{ii} = 0$, $\forall i \in V$. It is important to underline that $M_{distances}$ is independent of the used vehicle. Based on this distance matrix and given a unit cost c^k (per kilometer) for vehicle k, a cost matrix M^k_{costs} is to be computed so that:

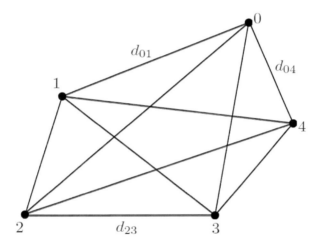

Figure 4.4 A graphic representation of the VL-DCVRP.

$$c^k_{ij} = c^k \times d_{ij}$$

(4.2)

Hence, n cost matrices $M^1_{costs}, \dots, M^n_{costs}$ are obtained for vehicles 1,..., n respectively. Each cost matrix M^k_{costs} is also a symmetric square matrix of order $m + 1$, as $c^k_{ij} = c^k_{ji}$ and $i, j = 0, \dots, m$. The cost matrix M^k_{costs}, for vehicle k, is described as in figure 4.6. A solution to the problem stated above should fulfill the following conditions:

$$
M_{distances} =
\begin{pmatrix}
 & D & C_1 & C_2 & \ldots & C_m \\
D & 0 & d_{01} & d_{02} & \ldots & d_{0m} \\
C_1 & d_{10} & 0 & d_{12} & \ldots & d_{1m} \\
C_2 & d_{20} & d_{21} & 0 & \ldots & d_{2m} \\
\vdots & & & \ddots & & \\
C_m & d_{m0} & d_{m1} & d_{m2} & \ldots & 0
\end{pmatrix}
$$

Figure 4.5 General form of the VL-DCVRP distance matrix.

$$
C^k =
\begin{pmatrix}
 & D & C_1 & C_2 & \ldots & C_m \\
D & 0 & c_{01}^k & c_{02}^k & \ldots & c_{0m}^k \\
C_1 & c_{10}^k & 0 & c_{12}^k & \ldots & c_{1m}^k \\
C_2 & c_{20}^k & c_{21}^k & 0 & \ldots & c_{2m}^k \\
\vdots & & & \ddots & & \\
C_m & c_{m0}^k & c_{m1}^k & c_{m2}^k & \ldots & 0
\end{pmatrix}
$$

Figure 4.6 General form of the VL-DCVRP cost matrix for vehicle k.

- The order of each customer j is assigned to exactly one of the n vehicles.
- The weight of the cargo packed into vehicle k must be in the range $[w^k_{min}, w^k_{max}]$ and should respect its capacity volume v^k_{max}.
- Each vehicle k can travel a total distance in the interval $[d^k_{min}, d^k_{max}]$.

Notation

We state in what follows the symbols used in the modeling of the VL-DCVRP (4.3)–(4.11):

Notation	Explanation
INPUTS	
n	The total number of vehicles
m	The total number of customers
V	The set of vertices representing the depot and customers
E	The set of edges expressing possible direct routes for each pair of vertices
v_i	The order's volume of customer i
w_i	The order's weight of customer i
c^k	The unit cost, per kilometer, for vehicle k
d_{ij}	The distance between customers i and j
$c_{ij}^k = c^k \times d_{ij}$	The cost of traveling between customers i and j using vehicle k
v_{max}^k	The volume capacity of vehicle k
$[w_{min}^k, w_{max}^k]$	The range of weight capacity that can be supported by vehicle k
$[d_{min}^k, d_{max}^k]$	The range of the total distance that vehicle k can travel
SE	The set of possible subours that can be built from V
OUPUTS	
$x_{ij}^k =$	$\begin{cases} 1 & \text{if the edge } \{i,j\} \text{ is traversed by vehicle } k \\ 0 & \text{elsewhere} \end{cases}$
$y^k =$	$\begin{cases} 1 & \text{if vehicle } k \text{ is used} \\ 0 & \text{elsewhere} \end{cases}$
DV	The set of decision variables including both x_{ij}^k and y^k

Mathematical model

The VL-DCVRP mathematical formulation is proposed as follows:

$$Min\ Z(x) = \sum_{k=1}^{n} \sum_{i=0}^{m} \sum_{i \neq j, j=0}^{m} c_{ij}^k\, x_{ij}^k \tag{4.3}$$

$$S.t. \quad \sum_{j=1}^{m} x_{0j}^k = \sum_{i=1}^{m} x_{i0}^k = 1, \ k = 1, \ldots, n \tag{4.4}$$

$$\sum_{k=1}^{n} \sum_{i \neq j, i=0}^{m} x_{ij}^{k} = 1, \ j = 1, \ldots, m \tag{4.5}$$

$$\sum_{i=0}^{m} x_{ij}^{k} - \sum_{i=0}^{m} x_{ji}^{k} = 0, \ j = 1, \ldots, m, \ k = 1, \ldots, n \tag{4.6}$$

$$\sum_{i \in S} \sum_{j \in S} x_{ij}^{k} \leq |S| - 1, \ k = 1, \ldots, n, \ \forall S \subseteq V, \ |S| \in \{2, \ldots, m\} \tag{4.7}$$

$$\sum_{i=0}^{m} \sum_{j \neq i, j=1}^{m} v_{j} \, x_{ij}^{k} \leq v_{max}^{k} \, y^{k}, \ k = 1, \ldots, n \tag{4.8}$$

$$w_{min}^{k} \, y^{k} \leq \sum_{i=0}^{m} \sum_{j \neq i, j=1}^{m} w_{j} \, x_{ij}^{k} \leq w_{max}^{k} \, y^{k}, \ k = 1, \ldots, n \tag{4.9}$$

$$d_{min}^{k} \, y^{k} \leq \sum_{i=0}^{m} \sum_{j \neq i, j=0}^{m} d_{ij} \, x_{ij}^{k} \leq d_{max}^{k} \, y^{k}, \ k = 1, \ldots, n \tag{4.10}$$

$$x_{ij}^{k}, y^{k} \in \{0, 1\}, \ i, j_{i \neq j} = 1, \ldots, m, \ k = 1, \ldots, n \tag{4.11}$$

- **The objective function**
  Equation (4.3) seeks to minimize the traveling cost.
- **Structural constraints**
  The whole set of structural constraints (4.4)–(4.11) express both of the loading and the routing requirements.
  Constraints (4.4)–(4.8) and (4.11) are related to the standard formulation of the CVRP. We propose to enrich this basic formulation by constraints (4.9) and (4.10) in order to formulate the CVRP with distance constraints while orders are evaluated regarding two independent scales: the weight, w_i, and the volume, v_i, giving rise to the VL-DCVRP.

  - **Circuit requirements**
    Constraints (4.4) express that each travel should begin and end at the depot.

  - **Visiting each customer once**
    Constraints (4.5) mean that each customer is visited only once, that is each order should be transported by only one vehicle.

  - **Path continuity**
    The set of constraints (4.6) guarantee the route continuity in which case the ending of an edge corresponds to the starting of the next edge.

- **Subtour elimination**

  Constraints (4.7) ensure that each subset of vertices $S \subseteq V$ cannot contain a cycle through all vertices forming it.

  For each subtour $S \subseteq V$, the number of constraints corresponds to the ways of selecting $|S| = i$ customers from the total number of customers m. This number is obviously C_m^i.

  As the problem includes n vehicles, the number of constraints of a subtour elimination with $|S| = i$ is $C_m^i \times n$.

Consequently, the number of constraints $|SE|$ related to the subtour elimination, for all $S \subseteq V$, is computed in the following way:

$$|SE| = \sum_{i=2}^{m} \underbrace{C_m^i \times n}_{|S|=i} \tag{4.12}$$

- **Volume constraints**

  Constraints (4.8) guarantee that the total orders volume loaded into a vehicle should not exceed its maximum volume.

- **Weight constraints**

  The capacity of a vehicle should be respected so that the freight's total weight has to be in an allowed range, as reported in constraints (4.9)

- **Distance constraints**

  Constraints (4.10) express that each vehicle is able to run in a range defined in advance.

The solution x that corresponds to the output of the problem contains all decision variables DV listed as a line vector so that:

$$DV = (x_{01}^1, \ldots, x_{ij}^k, \ldots, x_{mm}^n, y^1, \ldots, y^n) \tag{4.13}$$

The number of binary decision variables DV is a summation of the number of x_{ij}^k and y^k, expressed as follows:

$$|DV| = \underbrace{\sum_{i=0}^{m}\sum_{j=1}^{m}\sum_{k=1}^{n}}_{|x_{ij}^k|} + \underbrace{\sum_{k=1}^{n}}_{|y^k|} = \underbrace{m \times n \times (m+1)}_{|x_{ij}^k|} + \underbrace{n}_{|y^k|} \tag{4.14}$$

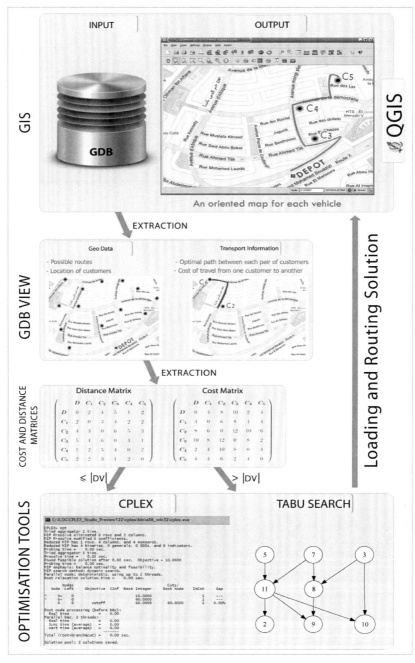

Figure 4.7 A GIS-O loose integration of QGIS and CPLEX-TS.

Color image of this figure appears in the color plate section at the end of the book.

Mathematically, the VRP variants shown in figure 4.2 are expressed as follows:

The CVRP: This version can be modeled mathematically by expressions from (4.3)–(4.8).

The DCVRP: Its formulation is expressed by (4.3)–(4.9).

The VL-CVRP: Its modeling corresponds to (4.3)–(4.8) and (4.10).

As the VL-DCVRP involves all structural constraints pointed out in figure 4.2, it is noticeable that compared to the above mentioned variations:

- Additional constraints increase the running time in solving the problem.
- As the branch and bound method needs an iterative branching until reaching the optimal solution, this number is proportional to the number of constraints.

4.4.3 Illustrations for the mathematical model

To clarify the use of the mathematical model (4.3)–(4.11) announced in the previous section, it is necessary to consider numerical examples and solve the problem at optimality using CPLEX. To this end, a series of VL-DCVRP instances are addressed in the present section to make all equations outlined earlier clearer.

A VL-DCVRP with $m = 5$ and $n = 2$

Let us consider a VL-DCVRP with ve orders and two vehicles. Each order is characterized by its volume and its weight. Each vehicle has a weight capacity that varies in $[w^k_{min}, w^k_{max}]$, a maximum volume v^k_{max} and a travel distance ranging in the interval $[d^k_{min}, d^k_{max}]$. Indeed, distance and cost matrices are reported in tables 4.8 and 4.9. All these settings are summarized in table 4.1.

Let us consider the following distance and cost matrices:

The input data reported in the distance matrix, the cost matrix and table 4.1 are formulated mathematically as follows:

Table 4.1 Inputs of the VL-DCVRP with $m = 5$ orders and $n = 2$.

$Vehicle^k$	1	2
c^k	2	2
$[w^k_{min}, w^k_{max}]$	[0,900]	[0,1000]
v^k_{max}	6	7
$[d^k_{min}, d^k_{max}]$	[0,100]	[0,90]

$Order_i$	1	2	3	4	5
w_i	110	75	50	250	150
v_i	0.5	1.5	1.4	1.73	0.56

$$M_{distances} = \begin{pmatrix} & D & C_1 & C_2 & C_3 & C_4 & C_5 \\ D & 0 & 5.23 & 4.20 & 8.23 & 4.56 & 12.67 \\ C_1 & 5.23 & 0 & 1.03 & 13.34 & 8.61 & 18.34 \\ C_2 & 4.20 & 1.03 & 0 & 12.30 & 7.64 & 17 \\ C_3 & 8.23 & 13.34 & 12.30 & 0 & 4.83 & 5.66 \\ C_4 & 4.56 & 8.61 & 7.64 & 4.83 & 0 & 10.62 \\ C_5 & 12.67 & 18.34 & 17 & 5.66 & 10.62 & 0 \end{pmatrix}$$

Figure 4.8 Distance matrix of a VL-DCVRP with $m = 5$ and $n = 2$.

$$M^k_{costs} = \begin{pmatrix} & D & C_1 & C_2 & C_3 & C_4 & C_5 \\ D & 0 & 5.75 & 4.62 & 9.05 & 5.01 & 13.94 \\ C_1 & 5.75 & 0 & 1.13 & 14.68 & 9.47 & 20.17 \\ C_2 & 4.62 & 1.13 & 0 & 13.53 & 8.41 & 18.7 \\ C_3 & 9.05 & 14.68 & 13.53 & 0 & 5.31 & 6.22 \\ C_4 & 5.01 & 9.47 & 8.41 & 5.31 & 0 & 11.69 \\ C_5 & 13.94 & 20.17 & 18.7 & 6.22 & 11.69 & 0 \end{pmatrix}$$

Figure 4.9 Cost matrix of a VL-DCVRP with $m = 5$ and $n = 2$.

- **The objective function (4.3)**

$$Min\ Z(x) =$$

$5.75\ x_{01}^1$	$+\ 4.62\ x_{02}^1$	$+\ 9.05\ x_{03}^1$	$+\ 5.01\ x_{04}^1$	$+\ 13.94\ x_{05}^1$	$+$
$5.75\ x_{10}^1$	$+\ 1.13\ x_{12}^1$	$+\ 14.68\ x_{13}^1$	$+\ 9.47\ x_{14}^1$	$+\ 20.17\ x_{15}^1$	$+$
$4.62\ x_{20}^1$	$+\ 1.13\ x_{21}^1$	$+\ 13.53\ x_{23}^1$	$+\ 8.41\ x_{24}^1$	$+\ 18.7\ x_{25}^1$	$+$
$9.05\ x_{30}^1$	$+\ 14.68\ x_{31}^1$	$+\ 13.53\ x_{32}^1$	$+\ 5.31\ x_{34}^1$	$+\ 6.22\ x_{35}^1$	$+$
$5.01\ x_{40}^1$	$+\ 9.47\ x_{41}^1$	$+\ 8.41\ x_{42}^1$	$+\ 5.31\ x_{43}^1$	$+\ 11.69\ x_{45}^1$	$+$
$13.94\ x_{50}^1$	$+\ 20.17\ x_{51}^1$	$+\ 18.7\ x_{52}^1$	$+\ 6.22\ x_{53}^1$	$+\ 11.69\ x_{54}^1$	$+$
$5.75\ x_{01}^2$	$+\ 4.62\ x_{02}^2$	$+\ 9.05\ x_{03}^2$	$+\ 5.01\ x_{04}^2$	$+\ 13.94\ x_{05}^2$	$+$
$5.75\ x_{10}^2$	$+\ 1.13\ x_{12}^2$	$+\ 14.68\ x_{13}^2$	$+\ 9.47\ x_{14}^2$	$+\ 20.17\ x_{15}^2$	$+$
$4.62\ x_{20}^2$	$+\ 1.13\ x_{21}^2$	$+\ 13.53\ x_{23}^2$	$+\ 8.41\ x_{24}^2$	$+\ 18.7\ x_{25}^2$	$+$
$9.05\ x_{30}^2$	$+\ 14.68\ x_{31}^2$	$+\ 13.53\ x_{32}^2$	$+\ 5.31\ x_{34}^2$	$+\ 6.22\ x_{35}^2$	$+$
$5.01\ x_{40}^2$	$+\ 9.47\ x_{41}^2$	$+\ 8.41\ x_{42}^2$	$+\ 5.31\ x_{43}^2$	$+\ 11.69\ x_{45}^2$	$+$
$13.94\ x_{50}^2$	$+\ 20.17\ x_{51}^2$	$+\ 18.7\ x_{52}^2$	$+\ 6.22\ x_{53}^2$	$+\ 11.69\ x_{54}^2$	

- **Circuit requirements (4.4)**

$$x_{01}^1 + x_{02}^1 + x_{03}^1 + x_{04}^1 + x_{05}^1 = 1$$
$$x_{01}^2 + x_{02}^2 + x_{03}^2 + x_{04}^2 + x_{05}^2 = 1$$
$$x_{10}^1 + x_{20}^1 + x_{30}^1 + x_{40}^1 + x_{50}^1 = 1$$
$$x_{10}^2 + x_{20}^2 + x_{30}^2 + x_{40}^2 + x_{50}^2 = 1$$

- **Visiting each customer once (4.5)**

$$x_{01}^1 + x_{21}^1 + x_{31}^1 + x_{41}^1 + x_{51}^1 + x_{01}^2 + x_{21}^2 + x_{31}^2 + x_{41}^2 + x_{51}^2 = 1$$
$$x_{02}^1 + x_{12}^1 + x_{32}^1 + x_{42}^1 + x_{52}^1 + x_{02}^2 + x_{12}^2 + x_{32}^2 + x_{42}^2 + x_{52}^2 = 1$$
$$x_{03}^1 + x_{13}^1 + x_{23}^1 + x_{43}^1 + x_{53}^1 + x_{03}^2 + x_{13}^2 + x_{23}^2 + x_{43}^2 + x_{53}^2 = 1$$
$$x_{04}^1 + x_{14}^1 + x_{24}^1 + x_{34}^1 + x_{54}^1 + x_{04}^2 + x_{14}^2 + x_{24}^2 + x_{34}^2 + x_{54}^2 = 1$$
$$x_{05}^1 + x_{15}^1 + x_{25}^1 + x_{35}^1 + x_{45}^1 + x_{05}^2 + x_{15}^2 + x_{25}^2 + x_{35}^2 + x_{45}^2 = 1$$

- **Path continuity (4.6)**

First vehicle : k = 1

$$x_{01}^1 + x_{21}^1 + x_{31}^1 + x_{41}^1 + x_{51}^1 - x_{10}^1 - x_{12}^1 - x_{13}^1 - x_{14}^1 - x_{15}^1 = 0$$

$$x_{02}^1 + x_{12}^1 + x_{32}^1 + x_{42}^1 + x_{52}^1 - x_{20}^1 - x_{21}^1 - x_{23}^1 - x_{24}^1 - x_{25}^1 = 0$$

$$x_{03}^1 + x_{13}^1 + x_{23}^1 + x_{43}^1 + x_{53}^1 - x_{30}^1 - x_{31}^1 - x_{32}^1 - x_{34}^1 - x_{35}^1 = 0$$

$$x_{04}^1 + x_{14}^1 + x_{24}^1 + x_{34}^1 + x_{54}^1 - x_{40}^1 - x_{41}^1 - x_{42}^1 - x_{43}^1 - x_{45}^1 = 0$$

$$x_{05}^1 + x_{15}^1 + x_{25}^1 + x_{35}^1 + x_{45}^1 - x_{50}^1 - x_{51}^1 - x_{52}^1 - x_{53}^1 - x_{54}^1 = 0$$

Second vehicle : k = 2

$$x_{01}^2 + x_{21}^2 + x_{31}^2 + x_{41}^2 + x_{51}^2 - x_{10}^2 - x_{12}^2 - x_{13}^2 - x_{14}^2 - x_{15}^2 = 0$$

$$x_{02}^2 + x_{12}^2 + x_{32}^2 + x_{42}^2 + x_{52}^2 - x_{20}^2 - x_{21}^2 - x_{23}^2 - x_{24}^2 - x_{25}^2 = 0$$

$$x_{03}^2 + x_{13}^2 + x_{23}^2 + x_{43}^2 + x_{53}^2 - x_{30}^2 - x_{31}^2 - x_{32}^2 - x_{34}^2 - x_{35}^2 = 0$$

$$x_{04}^2 + x_{14}^2 + x_{24}^2 + x_{34}^2 + x_{54}^2 - x_{40}^2 - x_{41}^2 - x_{42}^2 - x_{43}^2 - x_{45}^2 = 0$$

$$x_{05}^2 + x_{15}^2 + x_{25}^2 + x_{35}^2 + x_{45}^2 - x_{50}^2 - x_{51}^2 - x_{52}^2 - x_{53}^2 - x_{54}^2 = 0$$

- **Subtour elimination (4.7)**
  Applying equation (4.12), the number of constraints related to the subtour elimination is:

$$|SE| = \sum_{i=2}^{5} C_5^i \times 2 = 20 + 20 + 10 + 2 = 52 \tag{4.15}$$

First vehicle : $k = 1$	*Second vehicle :* $k = 2$
$x_{12}^1 + x_{21}^1 \leq 1$	$x_{12}^2 + x_{21}^2 \leq 1$
$x_{13}^1 + x_{31}^1 \leq 1$	$x_{13}^2 + x_{31}^2 \leq 1$
$x_{23}^1 + x_{32}^1 \leq 1$	$x_{23}^2 + x_{32}^2 \leq 1$
$x_{14}^1 + x_{41}^1 \leq 1$	$x_{14}^2 + x_{41}^2 \leq 1$
$x_{24}^1 + x_{42}^1 \leq 1$	$x_{24}^2 + x_{42}^2 \leq 1$
$x_{34}^1 + x_{43}^1 \leq 1$	$x_{34}^2 + x_{43}^2 \leq 1$
$x_{15}^1 + x_{51}^1 \leq 1$	$x_{15}^2 + x_{51}^2 \leq 1$
$x_{25}^1 + x_{52}^1 \leq 1$	$x_{25}^2 + x_{52}^2 \leq 1$
$x_{35}^1 + x_{53}^1 \leq 1$	$x_{35}^2 + x_{53}^2 \leq 1$
$x_{45}^1 + x_{54}^1 \leq 1$	$x_{45}^2 + x_{54}^2 \leq 1$

First vehicle : $k = 1$	*Second vehicle :* $k = 2$
$x_{12}^1 + x_{13}^1 + x_{21}^1 + x_{23}^1 + x_{31}^1 + x_{32}^1 \leq 2$	$x_{12}^2 + x_{13}^2 + x_{21}^2 + x_{23}^2 + x_{31}^2 + x_{32}^2 \leq 2$
$x_{12}^1 + x_{14}^1 + x_{21}^1 + x_{24}^1 + x_{41}^1 + x_{42}^1 \leq 2$	$x_{12}^2 + x_{14}^2 + x_{21}^2 + x_{24}^2 + x_{41}^2 + x_{42}^2 \leq 2$
$x_{12}^1 + x_{15}^1 + x_{21}^1 + x_{25}^1 + x_{51}^1 + x_{52}^1 \leq 2$	$x_{12}^2 + x_{15}^2 + x_{21}^2 + x_{25}^2 + x_{51}^2 + x_{52}^2 \leq 2$
$x_{13}^1 + x_{14}^1 + x_{31}^1 + x_{34}^1 + x_{41}^1 + x_{43}^1 \leq 2$	$x_{13}^2 + x_{14}^2 + x_{31}^2 + x_{34}^2 + x_{41}^2 + x_{43}^2 \leq 2$
$x_{13}^1 + x_{15}^1 + x_{31}^1 + x_{35}^1 + x_{51}^1 + x_{53}^1 \leq 2$	$x_{13}^2 + x_{15}^2 + x_{31}^2 + x_{35}^2 + x_{51}^2 + x_{53}^2 \leq 2$
$x_{14}^1 + x_{15}^1 + x_{41}^1 + x_{45}^1 + x_{51}^1 + x_{54}^1 \leq 2$	$x_{14}^2 + x_{15}^2 + x_{41}^2 + x_{45}^2 + x_{51}^2 + x_{54}^2 \leq 2$
$x_{23}^1 + x_{24}^1 + x_{32}^1 + x_{34}^1 + x_{42}^1 + x_{43}^1 \leq 2$	$x_{23}^2 + x_{24}^2 + x_{32}^2 + x_{34}^2 + x_{42}^2 + x_{43}^2 \leq 2$
$x_{23}^1 + x_{25}^1 + x_{32}^1 + x_{35}^1 + x_{52}^1 + x_{53}^1 \leq 2$	$x_{23}^2 + x_{25}^2 + x_{32}^2 + x_{35}^2 + x_{52}^2 + x_{53}^2 \leq 2$
$x_{24}^1 + x_{25}^1 + x_{42}^1 + x_{45}^1 + x_{52}^1 + x_{54}^1 \leq 2$	$x_{24}^2 + x_{25}^2 + x_{42}^2 + x_{45}^2 + x_{52}^2 + x_{54}^2 \leq 2$
$x_{43}^1 + x_{45}^1 + x_{34}^1 + x_{35}^1 + x_{54}^1 + x_{53}^1 \leq 2$	$x_{43}^2 + x_{45}^2 + x_{34}^2 + x_{35}^2 + x_{54}^2 + x_{53}^2 \leq 2$

First vehicle : k = 1

$$x^1_{12} + x^1_{13} + x^1_{14} + x^1_{21} + x^1_{23} + x^1_{24} + x^1_{31} + x^1_{32} + x^1_{34} + x^1_{41} + x^1_{42} + x^1_{43} \leq 3$$

$$x^1_{12} + x^1_{13} + x^1_{15} + x^1_{21} + x^1_{23} + x^1_{25} + x^1_{31} + x^1_{32} + x^1_{35} + x^1_{51} + x^1_{52} + x^1_{53} \leq 3$$

$$x^1_{12} + x^1_{14} + x^1_{15} + x^1_{21} + x^1_{24} + x^1_{25} + x^1_{41} + x^1_{42} + x^1_{45} + x^1_{51} + x^1_{52} + x^1_{54} \leq 3$$

$$x^1_{13} + x^1_{14} + x^1_{15} + x^1_{31} + x^1_{34} + x^1_{35} + x^1_{41} + x^1_{43} + x^1_{45} + x^1_{51} + x^1_{53} + x^1_{54} \leq 3$$

$$x^1_{23} + x^1_{24} + x^1_{25} + x^1_{32} + x^1_{34} + x^1_{35} + x^1_{42} + x^1_{43} + x^1_{45} + x^1_{52} + x^1_{53} + x^1_{54} \leq 3$$

Second vehicle : k = 2

$$x^2_{12} + x^2_{13} + x^2_{14} + x^2_{21} + x^2_{23} + x^2_{24} + x^2_{31} + x^2_{32} + x^2_{34} + x^2_{41} + x^2_{42} + x^2_{43} \leq 3$$

$$x^2_{12} + x^2_{13} + x^2_{15} + x^2_{21} + x^2_{23} + x^2_{25} + x^2_{31} + x^2_{32} + x^2_{35} + x^2_{51} + x^2_{52} + x^2_{53} \leq 3$$

$$x^2_{12} + x^2_{14} + x^2_{15} + x^2_{21} + x^2_{24} + x^2_{25} + x^2_{41} + x^2_{42} + x^2_{45} + x^2_{51} + x^2_{52} + x^2_{54} \leq 3$$

$$x^2_{13} + x^2_{14} + x^2_{15} + x^2_{31} + x^2_{34} + x^2_{35} + x^2_{41} + x^2_{43} + x^2_{45} + x^2_{51} + x^2_{53} + x^2_{54} \leq 3$$

$$x^2_{23} + x^2_{24} + x^2_{25} + x^2_{32} + x^2_{34} + x^2_{35} + x^2_{42} + x^2_{43} + x^2_{45} + x^2_{52} + x^2_{53} + x^2_{54} \leq 3$$

First vehicle : k = 1

$$x^1_{21} + x^1_{31} + x^1_{41} + x^1_{51} + x^1_{12} + x^1_{32} + x^1_{42} + x^1_{52} + x^1_{13} + x^1_{23} + x^1_{43} + x^1_{53}$$
$$+ x^1_{14} + x^1_{24} + x^1_{34} + x^1_{54} + x^1_{15} + x^1_{25} + x^1_{35} + x^1_{45} \leq 4$$

Second vehicle : k = 2

$$x^2_{21} + x^2_{31} + x^2_{41} + x^2_{51} + x^2_{12} + x^2_{32} + x^2_{42} + x^2_{52} + x^2_{13} + x^2_{23} + x^2_{43} + x^2_{53}$$
$$+ x^2_{14} + x^2_{24} + x^2_{34} + x^2_{54} + x^2_{15} + x^2_{25} + x^2_{35} + x^2_{45} \leq 4$$

- **Volume constraints (4.8)**

First vehicle : k = 1

$$0.5x^1_{01} + 1.5x^1_{02} + 1.4x^1_{03} + 1.73x^1_{04} + 0.56x^1_{05} + 1.5x^1_{12} + 1.4x^1_{13} + 1.73x^1_{14} +$$
$$0.56x^1_{15} + 0.5x^1_{21} + 1.4x^1_{23} + 1.73x^1_{24} + 0.56x^1_{25} + 0.5x^1_{31} + 1.5x^1_{32} + 1.73x^1_{34} +$$
$$0.56x^1_{35} + 0.5x^1_{41} + 1.5x^1_{42} + 1.4x^1_{43} + 0.56x^1_{45} + 0.5x^1_{51} + 1.5x^1_{52} + 1.4x^1_{53} +$$
$$1.73x^1_{54} - 6y^1 \leq 0$$

Second vehicle : $k = 2$

$$0.5x_{01}^2 + 1.5x_{02}^2 + 1.4x_{03}^2 + 1.73x_{04}^2 + 0.56x_{05}^2 + 1.5x_{12}^2 + 1.4x_{13}^2 + 1.73x_{14}^2 +$$
$$0.56x_{15}^2 + 0.5x_{21}^2 + 1.4x_{23}^2 + 1.73x_{24}^2 + 0.56x_{25}^2 + 0.5x_{31}^2 + 1.5x_{32}^2 + 1.73x_{34}^2 +$$
$$0.56x_{35}^2 + 0.5x_{41}^2 + 1.5x_{42}^2 + 1.4x_{43}^2 + 0.56x_{45}^2 + 0.5x_{51}^2 + 1.5x_{52}^2 + 1.4x_{53}^2 +$$
$$1.73x_{54}^2 - 7y^2 \le 0$$

• Weight constraints (4.9)

First vehicle : $k = 1$

$$110x_{01}^1 + 75x_{02}^1 + 50x_{03}^1 + 250x_{04}^1 + 150x_{05}^1 + 75x_{12}^1 + 50x_{13}^1 + 250x_{14}^1 + 150x_{15}^1 +$$
$$110x_{21}^1 + 50x_{23}^1 + 250x_{24}^1 + 150x_{25}^1 + 110x_{31}^1 + 75x_{32}^1 + 250x_{34}^1 + 150x_{35}^1 + 110x_{41}^1 +$$
$$75x_{42}^1 + 50x_{43}^1 + 150x_{45}^1 + 110x_{51}^1 + 75x_{52}^1 + 50x_{53}^1 + 250x_{54}^1 - 900y^1 \le 0$$

Second vehicle : $k = 2$

$$110x_{01}^2 + 75x_{02}^2 + 50x_{03}^2 + 250x_{04}^2 + 150x_{05}^2 + 75x_{12}^2 + 50x_{13}^2 + 250x_{14}^2 + 150x_{15}^2 +$$
$$110x_{21}^2 + 50x_{23}^2 + 250x_{24}^2 + 150x_{25}^2 + 110x_{31}^2 + 75x_{32}^2 + 250x_{34}^2 + 150x_{35}^2 + 110x_{41}^2 +$$
$$75x_{42}^2 + 50x_{43}^2 + 150x_{45}^2 + 110x_{51}^2 + 75x_{52}^2 + 50x_{53}^2 + 250x_{54}^2 - 1000y^2 \le 0$$

• Distance constraints (4.10)

First vehicle : $k = 1$

$$5.23x_{01}^1 + 4.20x_{02}^1 + 8.23x_{03}^1 + 4.56x_{04}^1 + 12.67x_{05}^1 + 1.03x_{12}^1 + 13.34x_{13}^1 + 8.61x_{14}^1 +$$
$$18.34x_{15}^1 + 1.03x_{21}^1 + 12.3x_{23}^1 + 7.64x_{24}^1 + 17x_{25}^1 + 13.34x_{31}^1 + 12.3x_{32}^1 + 4.83x_{34}^1 +$$
$$5.66x_{35}^1 + 8.61x_{41}^1 + 7.64x_{42}^1 + 4.83x_{43}^1 + 10.62x_{45}^1 + 18.34x_{51}^1 + 17x_{52}^1 + 5.66x_{53}^1 +$$
$$10.62x_{54}^1 + 5.23x_{10}^1 + 4.2x_{20}^1 + 8.23x_{30}^1 + 4.56x_{40}^1 + 12.67x_{50}^1 - 100y^1 \le 0$$

$$-5.23x_{01}^1 - 4.2x_{02}^1 - 8.23x_{03}^1 - 4.56x_{04}^1 - 12.67x_{05}^1 - 1.03x_{12}^1 - 13.34x_{13}^1 - 8.61x_{14}^1 -$$
$$18.34x_{15}^1 - 1.03x_{21}^1 - 12.3x_{23}^1 - 7.64x_{24}^1 - 17x_{25}^1 - 13.34x_{31}^1 - 12.3x_{32}^1 - 4.83x_{34}^1 -$$
$$5.66x_{35}^1 - 8.61x_{41}^1 - 7.64x_{42}^1 - 4.83x_{43}^1 - 10.62x_{45}^1 - 18.34x_{51}^1 - 17x_{52}^1 - 5.66x_{53}^1 -$$
$$10.62x_{54}^1 - 5.23x_{10}^1 - 4.2x_{20}^1 - 8.23x_{30}^1 - 4.56x_{40}^1 - 12.67x_{50}^1 \le 0$$

Second vehicle : $k = 2$

$5.23x_{01}^2 + 4.20x_{02}^2 + 8.23x_{03}^2 + 4.56x_{04}^2 + 12.67x_{05}^2 + 1.03x_{12}^2 + 13.34x_{13}^2 + 8.61x_{14}^2$

$+18.34x_{15}^2 + 1.03x_{21}^2 + 12.3x_{23}^2 + 7.64x_{24}^2 + 17x_{25}^2 + 13.34x_{31}^2 + 12.3x_{32}^2 + 4.83x_{34}^2 +$

$5.66x_{35}^2 + 8.61x_{41}^2 + 7.64x_{42}^2 + 4.83x_{43}^2 + 10.62x_{45}^2 + 18.34x_{51}^2 + 17x_{52}^2 + 5.66x_{53}^2 +$

$10.62x_{54}^2 + 5.23x_{10}^2 + 4.2x_{20}^2 + 8.23x_{30}^2 + 4.56x_{40}^2 + 12.67x_{50}^2 - 90y^2 \leq 0$

$-5.23x_{01}^2 - 4.2x_{02}^2 - 8.23x_{03}^2 - 4.56x_{04}^2 - 12.67x_{05}^2 - 1.03x_{12}^2 - 13.34x_{13}^2 - 8.61x_{14}^2 -$

$18.34x_{15}^2 - 1.03x_{21}^2 - 12.3x_{23}^2 - 7.64x_{24}^2 - 17x_{25}^2 - 13.34x_{31}^2 - 12.3x_{32}^2 - 4.83x_{34}^2 -$

$5.66x_{35}^2 - 8.61x_{41}^2 - 7.64x_{42}^2 - 4.83x_{43}^2 - 10.62x_{45}^2 - 18.34x_{51}^2 - 17x_{52}^2 - 5.66x_{53}^2$

$-10.62x_{54}^2 - 5.23x_{10}^2 - 4.2x_{20}^2 - 8.23x_{30}^2 - 4.56x_{40}^2 - 12.67x_{50}^2 \leq 0$

CPLEX is applied to solve the above formulation, as shown in figure 4.10. The output is a description of the objective function value, the elapsed time and the solution details (DV). By equation (4.14), the number of decision variables is:

$$|DV| = 5 \times 2 \times 6 + 2 = 62 \qquad (4.16)$$

Figure 4.10 Solution details by CPLEX for $m = 5$ and $n = 2$.

Applying CPLEX to solve VRP variants announced so far with $m = 5$ and $n = 2$ (CVRP, DCVRP, VL-CVRP and VL-DCVRP), all problems have the same optimal solution and require the same running time. This result is expected as the problem is small-sized. For a more complete analysis of the resources needed to solve all VRP variants, we address in the next part a VL-DCVRP with $m = 9$ customers and $n = 3$ vehicles.

A VL-DCVRP with $m = 9$ and $n = 3$

We propose to solve in this part the case $m = 9$ and $n = 3$ giving rise to the following number of decision variables:

$$|DV| = 9 \times 3 \times 10 + 3 = 273 \tag{4.17}$$

If we proceed to a computation of the number of constraints related to the subtour elimination, it follows that:

$$|SE| = \sum_{i=2}^{9} C_9^i \times 3 = 502 \tag{4.18}$$

The input data for the distance matrix $M_{distances}$ and cost matrices M^k_{costs} are reported in figures 4.11 and 4.12.

The data related to the configuration of each vehicle k for $k = 1,2,3$, the weight w_i and the volume v_i of each customer i are reported in table 4.2 for $i = 1,...,9$.

As the number of constraints is too large, their exhaustive enumeration is useless and heavy to read. Therefore, we simply report the final solution generated by CPLEX in figure 4.13. This figure reports solution details for the four variations. The second column of table 4.3 prints out the time, in seconds, required to solve the related variation. We can observe that the VL-DCVRP is the most time consuming due to the number of constraints that it contains. The third column of table 4.3 describes the vehicles pathways which cost 126.59 for all VRP variants.

$$M_{distances} =$$

	D	C_1	C_2	C_3	C_4	C_5	C_6	C_7	C_8	C_9
D	0	5.23	4.20	8.23	4.56	12.67	15.21	19.55	21.89	54.07
C_1	5.23	0	1.03	13.34	8.61	18.34	19.77	26.09	23.86	48.83
C_2	4.20	1.03	0	12.30	7.64	17	19.47	26.04	23.69	47.80
C_3	8.23	13.34	12.30	0	4.83	5.66	7.09	13.87	11.51	35.50
C_4	4.56	8.61	7.64	4.83	0	10.62	11.95	18.65	16.24	30.67
C_5	12.67	18.34	17	5.66	10.62	0	2.56	9.21	6.87	20.04
C_6	15.21	19.77	19.47	7.09	11.95	2.56	0	6.71	4.38	17.63
C_7	19.55	26.09	26.04	13.87	18.65	9.21	6.71	0	2.35	10.95
C_8	21.89	23.86	23.69	11.51	16.24	6.87	4.38	2.35	0	13.29
C_9	54.07	48.83	47.80	35.50	30.67	20.04	17.63	10.95	13.29	0

Figure 4.11 Distance matrix of a VL-DCVRP with $m = 9$ and $n = 3$.

$$M^k_{costs} =$$

	D	C_1	C_2	C_3	C_4	C_5	C_6	C_7	C_8	C_9
D	0	5.75	4.62	9.05	5.01	13.94	16.73	21.50	24.08	59.47
C_1	5.75	0	1.13	14.68	9.47	20.17	21.75	28.70	26.25	53.72
C_2	4.62	1.13	0	13.53	8.41	18.7	21.42	28.65	26.06	52.58
C_3	9.05	14.68	13.53	0	5.31	6.22	7.80	15.26	12.66	39.05
C_4	5.01	9.47	8.41	5.31	0	11.69	13.14	20.51	17.86	33.73
C_5	13.94	20.17	18.7	6.22	11.69	0	2.81	10.13	7.56	22.04
C_6	16.73	21.75	21.42	7.80	13.14	2.81	0	7.38	4.81	19.39
C_7	21.50	28.70	28.65	15.26	20.51	10.13	7.38	0	2.58	12.05
C_8	24.08	26.25	26.06	12.66	17.86	7.56	4.81	2.58	0	14.62
C_9	59.47	53.72	52.58	39.05	33.73	22.04	19.39	12.05	14.62	0

Figure 4.12 Cost matrix of a VL-DCVRP with $m = 9$ and $n = 3$.

Table 4.2 Inputs of the VL-DCVRP with $m = 9$ and $n = 3$.

$Vehicle^k$	1	2	3
c^k	2	2	2
$[w^k_{min}, w^k_{max}]$	[0,900]	[0,1000]	[0,900]
v^k_{max}	6	7	5
$[d^k_{min}, d^k_{max}]$	[0,100]	[0,90]	[0,40]

$Order_i$	1	2	3	4	5	6	7	8	9
w_i	110	75	50	250	150	75	125	250	385
v_i	0.5	1.5	1.4	1.73	0.56	0.16	0.5	1	1.35

4.4.4 Results by CPLEX for $5 \leq m \leq 9$ and $2 \leq n \leq 3$

To generalize the above discussion, we solve a series of VL-DCVRP instances varying from $m = 5$ and $n = 2$ to $m = 9$ and $n = 3$. As m can take 5 different values ($9-5 + 1 = 5$) and n takes 2 values, $5 \times 2 = 10$ VL-DCVRPs arise.

For each problem configuration, let i describe the value of m and k the value of n. We adopt the notation m_i-n_k to designate a VL-DCVRP with i customers and k vehicles. Table 4.4 reports, for each problem configuration,

Figure 4.13 Solution details by CPLEX for $m = 9$ and $n = 3$.

Table 4.3 CPLEX solutions for VRP variants with $m = 9$ and $n = 3$.

Problem	Time	Solution
CVRP	0.81	**(V1)** D → c3 → c4 → D
		(V2) D → c7 → c9 → c8 → c6 → c5 → D
		(V3) D → c1 → c2 → D
DCVRP	0.75	**(V1)** D → c3 → c4 → D
		(V2) D → c7 → c9 → c8 → c6 → c5 → D
		(V3) D → c2 → c1 → D
VL-CVRP	0.80	**(V1)** D → c3 → c4 → D
		(V2) D → c5 → c6 → c8 → c9 → c7 → D
		(V3) D → c1 → c2 → D
VL-DCVRP	0.86	**(V1)** D → c2 → c1 → D
		(V2) D → c5 → c6 → c8 → c9 → c7 → D
		(V3) D → c3 → c4 → D

the number of decision variables $|DV|$, the elapsed time for optimally solving the problem and the itinerary for each vehicle starting from and ending to the depot. We can observe from table 4.4:

- The progress of the CPU time over the number of decision variables $|DV|$ as reported in figure 4.14.
- The deterioration of the total cost from 42:01 to 126:59 due mainly to the enlargement of the number of customers m.

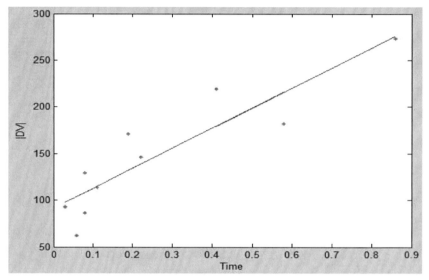

Figure 4.14 Variations of the CPU time in terms of $|DV|$.

Based on the generated results in table 4.4, MATLAB is applied to generate the following regression model:

$$Time = 38 \ 10 \times^{-4} \ |DV| - 0:29 \tag{4.19}$$

Equation (4.19) plotted in figure 4.14 allows, without further experiments, to predict the CPU time in terms of the problem size $|DV|$. In fact, for a VL-DCVRP with $m = 400$ and $n = 100$, the number of decision variables is $|DV| = 16\,040\,100$, as reported in equation (4.14). By equation (4.19), the time required to solve such large-scaled problem is 15 hours. However, when solving the problem by an approximate method as TS, the running time is about 20 minutes versus a tiny deviation from the optimal solution.

This illustration is an incentive to address real-world applications using an alternative framework that offers more flexibility for solving both small and large-scaled instances. The solution approach to be adopted can

Table 4.4 Results by CPLEX for $5 \leq m \leq 9$ and $2 \leq n \leq 3$.

Instance	|DV|	$Z(x^*)$	Time	Solution
$m_5\_n_2$	62	42.01	0.06	(V1) D → c4 → c3 → c5 → D
				(V2) D → c2 → c1 → D
$m_5\_n_3$	93	50.77	0.03	(V1) D → c5 → c3 → D
				(V2) D → c2 → c1 → D
				(V3) D → c4 → D
$m_6\_n_2$	86	46.40	0.08	(V1) D → c4 → c3 → c6 → c5 → D
				(V2) D → c2 → c1 → D
$m_6\_n_3$	129	55.16	0.08	(V1) D → c5 → c6 → c3 → D
				(V2) D → c2 → c1 → D
				(V3) D → c4 → D
$m_7\_n_2$	114	59.78	0.11	(V1) D → c2 → c1 → D
				(V2) D → c4 → c3 → c5 → c6 → c7 → D
$m_7\_n_3$	171	68.54	0.19	(V1) D → c3 → c5 → c6 → c7 → D
				(V2) D → c4 → D
				(V3) D → c2 → c1 → D
$m_8\_n_2$	146	59.79	0.22	(V1) D → c2 → c1 → D
				(V2) D → c4 → c3 → c5 → c6 → c8 → c7 → D
$m_8\_n_3$	219	68.55	0.41	(V1) D → c4 → D
				(V2) D → c3 → c5 → c6 → c8 → c7 → D
				(V3) D → c2 → c1 → D
$m_9\_n_2$	182	125.30	0.58	(V1) D → c2 → c1 → c4 → c3 → D
				(V2) D → c5 → c6 → c8 → c7 → c9 → D
$m_9\_n_3$	273	126.59	0.86	(V1) D → c2 → c1 → D
				(V2) D → c5 → c6 → c8 → c7 → c9 → D
				(V3) D → c3 → c4 → D

either be exact, as it is the case of the CPLEX optimizer, or approximate, as the TS metaheuristic depends on the problem size $|DV|$. To be more explicit in solving and showing solution details, we develop a GIS-O integration framework that addresses more suitably the VL-DCVRP by extracting real data from a specific area, solving the problem and visualizing cartographically the obtained solution.

4.5 A loose GIS-O integration for the VL-DCVRP

For an efficient resolution of the VL-DCVRP, we develop an environment that operationalizes a GIS-O integration characterized by a cost saving, a flexibility, an extensibility and a genericity that ensures the autonomy as well as the exchangeability of the involved sub-systems. These features allow changing the GIS, if required, and the optimization sub-system when the problem needs alternative solution approaches.

Our comparative study between integration strategies, developed in chapter 3, reveals that the loose integration strategy is the most appropriate design to be adopted for transportation problems.

Based on the above mentioned arguments, we adopt the loose coupling approach. We show firstly the architecture of the general model, then materialize the obtained framework by applying QGIS and CPLEX-TS.

4.5.1 General outline of the loose GIS-O integration

The general architecture of the loose integration between the GIS and an optimization system for solving the VL-DCVRP generates alternative loading and routing solutions geographically displayed to show their relevance with regards to real-time observed data.

Such dynamic data can be related to the state of the roads and the traffic in terms of the time interval for the delivery process. Other spatial data can be relevant in the loading and routing process as the neighborhood, the weather and social events.

We propose a loose coupling approach that consists in integrating a GIS with optimization tools. The output of this integration yields to a framework that maintains each sub-system apart and manages the data exchange process by macro languages.

As reported in figure 4.7, routing data are extracted from the GDB, then translated in distance and cost matrices. Those data are considered as inputs for the optimization step, using either an exact or an approximate approach, to generate a solution for the VL-DCVRP, depending on the problem size.

Each obtained solution is geographically visualized in the following way: for each vehicle a path is highlighted in a separate map followed with its loading configuration, as illustrated in figure 4.3. The loose integration for the VL-DCVRP can be described through steps of algorithm (9). According to this algorithm, the use of the loose coupling approach has to be detailed for each sub-system as indicated below:

STEP 1. Data extraction (GIS part)

The data extraction protocol depends on the structure of the GDB and the GIS to be adopted. Various GIS can be of interest as QGIS or OpenJUMP.

STEP 2. Matrices construction (GIS part)

Once these data are available, distance and cost matrices are built and capacity configurations of the vehicles are also provided to constitute the inputs of the optimization part.

Algorithm 9: The loose integration algorithm for the VL-DCVRP

STEP 1. Data extraction

Extract geographical data and transport information from the GIS

STEP 2. Matrices construction

Construct distance and cost matrices

STEP 3. Data transfer (GIS→OPTIMIZATION)

Transfer the dataset to the optimization component

STEP 4. Resolution

Apply the suitable tool within the optimization sub-system to generate a loading and routing solution

while (the DM is not satisfied with the generated solution) **do**
Apply the optimization tool depending on the problem size $|DV|$

end while

STEP 5. Data transfer (OPTIMIZATION→GIS)

Transfer the obtained solution to the GIS

STEP 6. Solution visualization

For each vehicle: provide its packing configuration and routing map

---

STEP 3. Data transfer (GIS part)

Transfer the dataset from the GIS to the optimization component.

STEP 4. Resolution (Optimization part)

The dataset is handled by the optimization sub-system that encompasses two main solvers applied depending on the problem size $|DV|$ as follows:

- $|DV| \leq th_{|DV|}$ If the the number of decision variables $|DV|$ is less than a prefixed threshold, say $th_{|DV|}$, then CPLEX, LINDO or GOROBI is applied to generate the optimal solution.
- $|DV| \leq th_{|DV|}$ If the number of decision variables exceeds $th_{|DV|}$, then an approximate approach is used. It can either be a heuristic specifically designed for the VL-DCVRP, as the greedy algorithm, or a metaheuristic as genetic algorithms (GAs), tabu search (TS), iterated local search (ILS) or simulated annealing (SA).

STEP 5. Data transfer (Optimization part)

Transfer the generated solution from the optimization component to the GIS.

STEP 6. Solution visualization (GIS part)

The obtained solution is displayed in a cartographic format.

4.5.2 A loose integration of QGIS and CPLEX-TS

To operationalize the loose integration for the VL-DCVRP, we propose a framework that performs GIS and optimization tools according to the loose integration strategy. QGIS software is performed for spatial data acquisition and visualization of the loading and routing solution. To solve the VL-DCVRP, CPLEX software is used as it is one of the most famous softwares that generates optimal solutions for small-sized problems. Being a NP-hard, the exact resolution for large instances of such problem is time consuming. Therefore, a metaheuristic, as the TS, is able to generate a promising solution in a reasonable computation time.

1. **GIS part:** We propose to perform Quantum GIS (QGIS) as it consists in an open source easily extensible for an integrated framework. QGIS can support numerous functionalities and formats and disposes of a user-friendly interface that manages factual, graphical and spatial queries as it permits an easy export of cartographical data.

   In our case, the extraction of routing data from QGIS is ensured by launching queries to the GDB. The obtained GDB-View that contains transport information is used to construct distance and cost matrices.

2. **Optimization part:** Two components are performed:

   - **CPLEX** as an optimizer that generates the optimal solution for limited sized VL-DCVRPs.
   - **Tabu search** as an approximate approach, that seems to be promising for routing problems.

This environment is a powerful tool for solving the VL-DCVRP efficiently while taking advantages from both of the above mentioned components.

The encoding of a solution x is described as follows:

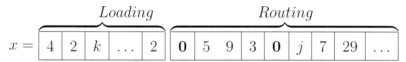

We can clearly see from the encoding of x that the vector is split up into two main components:

- **Loading:** Each position i in the loading part corresponds to the vehicle in which it is loaded. For example, orders 1, 2 and 3 are loaded in vehicles 4, 2 and k respectively.
- **Routing:** Depending on the stowage of each vehicle, the solution x describes the itinerary adopted by each vehicle where the departure and the termination are indexed by "0" referring to the depot.

We detail in algorithm (11) the main steps of the TS so far described in algorithm (10).

Algorithm 10: The loose integration of QGIS and CPLEX-TS

STEP 1. Data extraction

Extract geographical data and transport information from the GDB of QGIS

STEP 2. Matrices construction

Built distance and cost matrices

STEP 3. Data transfer (QGIS → CPLEX-TS)

Transfer the dataset to the optimization component

STEP 4. Resolution

if (the number of decision variables $|DV| \leq th_{|DV|}$) **then**

    Use the CPLEX solver to generate the optimal solution

else

    Apply the tabu search algorithm to get a near optimal solution

end if

STEP 5. Data transfer (CPLEX-TS → QGIS)

Transmit the obtained solution to QGIS

STEP 6. Solution visualization

Display, for each vehicle, its packing configuration and routing map

---

4.5.3 An illustrative example

We reconsider the VL-DCVRP with $m = 9$ and $n = 3$, already solved mathematically in section (4.4.3), and we apply the GIS-O loose coupling approach using QGIS and CPLEX-TS.

To materialize the GIS-O loose integration stated in algorithm (10), we apply it on Ezzahra town, located in the north of Tunisia, as shown in figure 4.15, and spread over 750 Hectars.

Based on the spatial data infrastructure of the addressed area, we localized the depot and the set of customers. The running of algorithm (10) follows the steps listed below:

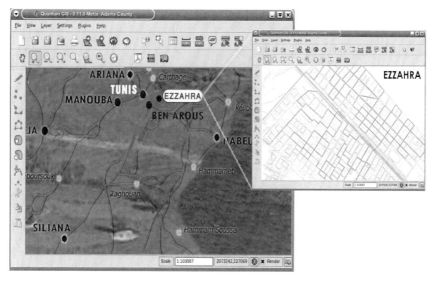

Figure 4.15 Location of the area under study: Ezzahra town.

Color image of this figure appears in the color plate section at the end of the book.

Algorithm 11: A tabu search approach for solving the VL-DCVRP

1: Generate an initial solution x_0 randomly or using the greedy algorithm

2: Evaluate its objective function $Z(x_0)$ and set the tabu search parameters

3: **while** (the termination condition is not satisfied) **do**

4:     Generate the neighborhood of the current solution x_c by simple permutations in the routing and loading configurations

5:     Select the best solution of the neighborhood as a current solution for the next iteration

6:     Check the tabu list

7: **end while**

8: Record the best encountered solution during the search process

---

- **STEP 1. Data extraction**
  The dataset of the problem as previously described in table 4.2 are:

$Order_i$	1	2	3	4	5	6	7	8	9
w_i	110	75	50	250	150	75	125	250	385
v_i	0.5	1.5	1.4	1.73	0.56	0.16	0.5	1	1.35

It is also assumed that vehicles k (k = 1,2,3) are set as follows:

- $c^k = 2$.
- $[w^k_{min}, w^k_{max}]$ is $[0, 6]$, $[0, 7]$ and $[0, 5]$.
- $[d^k_{min}, d^k_{max}]$ is $[0, 900]$, $[0, 1000]$ and $[0, 900]$.
- v^k_{max} is 100, 90 and 40.

- **STEP 2. Matrices construction**
  Based on the distance matrix $M_{distances}$ recorded in figure 4.11, the cost matrix M^k_{costs} is computed in such a way that each component $c^k_{ij} = c^k \times d_{ij}$. As $c^k = 2$ is the same for all vehicles,

  $$M^1_{costs} = M^2_{costs} = M^3_{costs}$$

  corresponding to figure 4.12.

- **STEP 3. Resolution**
  As outlined in algorithm (10), the solution of the VL-DCVRP can be generated using either the CPLEX solver, to get the optimal solution or TS stated in algorithm (11) that provides a near optimal solution in a reasonable computation time. This resolution is conditioned by the number of decision variables $|DV|$ under study. If we proceed by the computation of the number of decision variables, it follows that:

$$|DV| = \sum_{k=1}^{n} + \sum_{i=0}^{m}\sum_{j=1}^{m}\sum_{k=1}^{n} = n + mn(m+1) = 3 + 9 \times 3 \times 10 = 273 \quad (4.20)$$

CPLEX, performed to solve the problem at optimality, generates the optimal solution x^* that corresponds to $Z(x^*) = 126.59$.

Tabu search To test the effectiveness of the proposed TS, we tried the algorithm (11) and obtained the same solution already generated by CPLEX. The encoding of the solution x^* is:

	Loading								*Routing*									
1	1	3	3	2	2	2	2	2	0	2	1	0	5	6	8	7	9	0	3	4

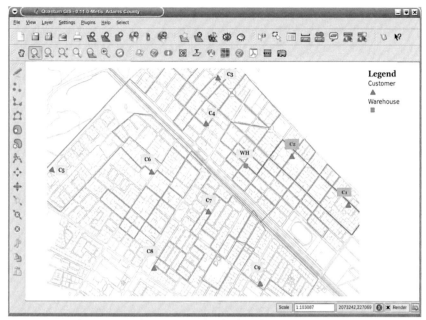

Figure 4.16 The routing of vehicle 1 for $m = 9$ and $n = 3$.
Color image of this figure appears in the color plate section at the end of the book.

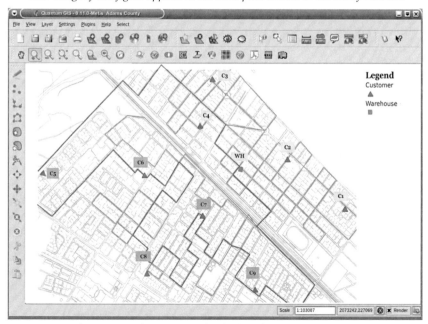

Figure 4.17 The routing of vehicle 2 for $m = 9$ and $n = 3$.
Color image of this figure appears in the color plate section at the end of the book.

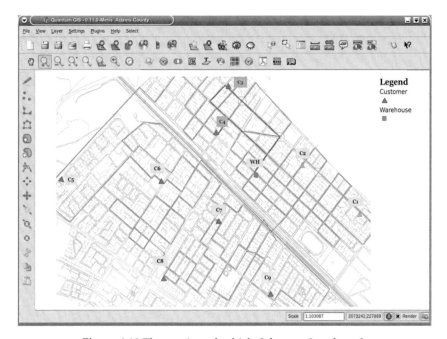

Figure 4.18 The routing of vehicle 3 for $m = 9$ and $n = 3$.

Color image of this figure appears in the color plate section at the end of the book.

4.6 Conclusion

We developed, in the present chapter, a framework for handling a fleet management problem specified by setting a loading configuration of items, then their delivery within a limited area. Given a set of items evaluated by their weights and volumes, a first stage consists in packing them into vehicles with prefixed configurations. Secondly, the design of the delivery takes place while minimizing the total cost. We stated the main components of this problem, known as the VL-DCVRP and its mathematical modeling while showing the limits of considering only the numerical side of the solution. Therefore, we proposed to solved the VL-DCVRP by tackling a GIS-O loose integration strategy that performs a GIS in one hand, and optimization tools in the other while exchanging information from one sub-system to another.

The optimization sub-system performs, depending on the problem size either an exact or an approximate approach. To try the GIS-O loose integration strategy, we applied QGIS for the GIS part and the CPLEX optimizer and the TS metaheuristic as the package for quantitatively solving the VL-DCVRP.

The study of some examples shows the effectiveness of the proposed approach and its accuracy from the data extraction to the visualization of the final solution in the map. This illustrates the adequacy of the generated solutions regarding real data and real maps.

Glossary

AP	:	Assignment Problem
CLP	:	Container Loading Problem
CVRP	:	Capacitated Vehicle Routing Problem
DLL	:	Dynamic Link Library
DM	:	Decision Maker
DBMS	:	Database Management Systems
ES	:	Efficient Solutions
GD	:	Geographical Data
GDB	:	Geographical DataBases
GIS	:	Geographical Information System
GIS-O	:	Integration of GIS and Optimization Tools
GUI	:	Graphical User Interface
LCVRP	:	Loading Capacitated Vehicle Routing Problem
OP	:	Optimization Problem
PSO	:	Particle Swarm Optimization
QGIS	:	Quantum GIS
SDSS	:	Spatial Decision Support System
SP	:	Scheduling Problem
TS	:	Tabu Search
TSP	:	traveling Salesman Problem
VL-DCVRP	:	Vector Loading Distance Capacitated Vehicle Routing Problem
3D	:	Three dimentional

References

Ai T.J. and Kachitvichyanukul V. (2009). Particle swarm optimization and two solution representations for solving the capacitated vehicle routing problem. Computers and Industrial Engineering. 56(1), 380–387.

Albrecht J. (1996). Geographic objects, and how to avoid them. In: Geographic objects with indeterminate boundaries. Burrough, P.A. and Frank, A.U. (Editors), Taylor and Francis, London. 325–331.

Alvarez M.D., Sans R., Garrido N. and Torres A. (2007). Factors that affect the quality of the bio-waste fraction of selectively collected solid waste in Catalonia. Waste Management. 28, 359–366.

Arampatzis, Kiranoudis G., Scaloubacas C.T.P. and Assimacopoulos, D. (2002). A GIS-based decision support system for planning urban transportation policies. European Journal of Operational Research. 152, 465–475.

Arentze T.A., Borgers A.W.J. and Timmerman H.J.P. (1996). Integrating GIS into the planning process. In: Spatial Analytical Perspectives 189 on GIS (Gisdata IV). Fischer M., Scholten H.J. and Unwin D. (Editors). Taylor and Francis, London. 187–198.

Awasthi A., Chauhan S., Parent M. and Proth J.M. (2011). Centralized fleet management system for cybernetic transportation. Expert Systems with Applications. 38, 3710–3717.

Batelaan O., De Smedt F. and Otero Valle M.N. (1993). Development and application of a groundwater model integrated in the GIS GRASS. Application of Geographic Information Systems in hydrology and water resources, The Netherlands. 581–589.

Batty M. and Xie, Y. (1994). Modeling inside GIS: Part 1. Model structures, exploratory spatial data analysis and aggregation. International Journal of Geographic Information Systems. 8(3), 291–307.

Beaulieu M. and Gamache M. (2006). An enumeration algorithm for solving the fleet management problem in underground mines. Computers and Operations Research. 33, 1606–1624.

Bellman R. (1953). An introduction to the theory of dynamic programming. Santa Monica, RAND Report. 109 pages.

Berry J.K., Buckley D.J. and McGarigal K. (1997). Seamlessly linking ARC/INFO and FRAGSTATS. Proceedings of Conference on Integrating Spatial Information technologies for Tomorrow, Vancouver. 413–416.

Bivand R. and Lucas A. (1997). Integrating models and geographical information systems. In: Geocomputation. Openshaw S. and Abrahart R.J. (Editors.). Taylor and Francis, London. 331–364.

Blazewicz J., Breit J., Formanowicz P., Kubiak W. and Schmidt G. (2001). Heuristic algorithms for the two-machine flowshop problem with limited machine availability. Omega. 29(6), 599-608.

Bodin L., Golden B., Assad A. and Ball M. (1983). Routing and scheduling of vehicles and crews: The state of the art. Computers and Operations Research. 10(2), 63–211.

Bowers K., Johnson S. and Pease K. (2004). Prospective hot-spotting—the future of crime mapping? British Journal of Criminology. 44(5), 641–658.

Bowers K.J., Johnson S.D. and Hirschfield A.F.G. (2004). Closing Off Opportunities for Crime: An Evaluation of Alley-Gating. European Journal on Criminal Policy and Research. 10(4), 285–308.

Brandmeyer J.E. and Karimi H.A. (2000). Coupling methodologies for environmental models. Environmental Modeling and Software. 15(5), 479–488.

Che C., Huang W., Lim A. and Zhu W. (2011). A heuristic for the multiple container loading cost minimization problem. In: Modern Approaches in Applied Intelligence. Mehrotra K., Mohan C., Oh J., Varshney P. and Ali M. (Editors). Lecture Notes in Computer Science 6704. Springer Verlag, Heidelberg. 276–285.

Chow V.T., Maidment D.R. and Mays L.W. (1988). Applied hydrology, New York: McGraw-Hill Publishing.p

Chrisman N. (1999). What does 'GIS' mean? Transactions in GIS. 3(2), 175–186.

Cordeau J.F., Gendreau M., Laporte G., Potvin J.Y. and Semet F. (2002). A guide to vehicle routing heuristics. Journal of the Operational Research Society. 53, 512–522.

Cornia G.C. and Riddell J. (2008). Toward a vision of Land in 2015. Lincoln Institute of Land Policy, Cambridge, Massachsetts. 335 pages.

Cowen D.J. (1988). GIS versus CAD versus DBMS: What are the Dierences? Photogrammetric, Engineering and Remote Sensing. 54, 1551–1555.

Cox A.B. and Gifford F., (1997). An overview to geographic information systems. The Journal of Academic Librarianship. 23(6), 449–461.

Craglia M., Haining R. and Wiles P. (2000). A comparative evaluation of approaches to urban crime pattern analysis. Urban Studies. 37, 711–729.

Curtin K.M., Hayslett-McCall K. and Qiu, F. (2007). Determining Optimal Police Patrol Areas with Maximal Covering and Backup Covering Location Models. 10, 125–145.

Dangermond J. (1982). A Classification of Software Components Used in GIS, American Cartography Association Conference, Denver, Colorado, Etats-Unis. 7–22.

Dangermond J. (1988). Trends in GIS and comments. Computers. Environment and Urban Systems. 12(3), 137–159.

Dantzig G.B (1956). Recent advances in linear programming. Management Science. 2, 131–144.

Dantzig G.B and Fulkerson D.R. (1954). Minimizing the number of Tankers to meet a fixed schedule. Naval Research Logistic Quartely. 1, 217–222.

DiBiase, D. (1990). Visualization in the Earth Sciences. Earth and Mineral Sciences. 59, 13–18.

Ding Y. and Fotheringham A.S. (1992). The integration of spatial analysis and GIS. Computers, Environment and Urban Systems. 16(1), 3–19.

Duhamel C., Lacomme P., Quilliot A. and Toussaint H. (2011). A multi-start evolutionary local search for the two-dimensional loading capacitated vehicle routing problem. Computers and Operations Research. 38, 617–640.

Dyckhoff H. (1990). A typology of cutting and packing problems. European Journal of Operational Research. 44(2), 145–159.

Egeblad J. and Pisinger D. (2009). Heuristic approaches for the two- and three-dimensional knapsack packing problem. Computers and Operations Research. 36(4), 1026–1049.

Esquirol P. and Lopez P. (1999). L'ordonnancement. Economica. Paris. 141 pages.

El-Kadi A.I., Oloufa A.A., Eltahan A.A. and Malik H.U. (1994). Use of a geographic information system in site-specific ground-water modelling. Ground Water. 32, 617–625.

Faiz S. (1999), Geographic information systems: Data quality and Data Mining, book in French, Ed. CLE, Tunis,Tunisia. 362 pages.

Faiz S. (2005). Knowledge discovery and geographical databases. Encyclopedia of Database Technologies and Applications. Rivero L., Doorn J. and Ferraggine V. (Editors). Idea Group, Etats-Unis. 308–311.

Faria G., Medeiros C.B. and Nascimento M.A. (1998). An extensible framework for spatio-temporal database applications. Proceedings of 194 10th International conference on. 202–205.

Fisher M.M. (1994). From conventional to knowledge-based geographic information systems. Computers, Environment and Urban Systems. 18(4), 233–242.

Fotheringham S. and Rogerson P. (1994). Spatial Analysis and GIS. Taylor and Francis. 281.

Gatrell A.C. (1987). On Putting Some Statistical Analysis into Geographical Information System with special reference to Problems of Map Comparison and Map Overlay.

Gendreau M., Potvin J.Y., Braysy O., Hasle G. and Lokketangen A. (2008). Metaheuristics for the vehicle routing problem and its extensions: A categorized bibliography. In The Vehicle Routing Problem: Latest Advances and New Challenges, Golden B., Raghavan S. and Wasil E. (Editors). Springer Verlag, Heidelberg. 143–169.

Gendreau M., Iori M., Laporte G. and Martello S. (2006). A tabu search algorithm for a routing and container loading problem. Transportation Science. 40(3), 342–350.

Ghose M.K., Dikshit A.K. and Sharma S.K. (2005). A GIS based transportation model for solid waste disposal—A case study on Asansol municipality. Waste Management. 26, 1287–1293.

Glover F. (1986). Future paths for integer programming and links to Artificial intelligence. Computers and Operations Research. 13, 533–549.

Gomes E.G. and Lins M.P.E. (2002). Integrating geographical information systems and multi-criteria methods: a case study. Annals of Operations Research. 116, 243–269.

Gorokhovich,Y. and Janus L. (1996). The NYC water quality division geographical information system (GIS) and its applications for the watershed management. Proceedings of Watershed'96 Technical Conference, Maryland. 530–532.

Goodchild M.F., Haining R. and Wise S. (1992). Integrating GIS and spatial data analysis: problems and possibilities. International Journal of Geographic Information Systems. 6(5), 407–423.

Grabaum R. and Meyer B.C. (1998). Multicriteria optimization of landscapes using GIS-based functional assessments. Landscape and Urban Planning. 43, 21–34.

Groenigen J.W., Stein A. and Zuurbier R. (1996). Optimization of environmental sampling using interactive GIS. Soil Technology. 10, 83–97.

Grubesic T.H. (2006). On the Application of Fuzzy Clustering for Crime Hot Spot Detection. Journal of Quantitative Criminology. 22(1), 77–105.

Haining R.P., Kerry R., Oliver M.A. (2010). Geography, Spatial Data Analysis, and Geostatistics: An Overview, Geographical Analysis Journal. 42(1), 7–31.

Harries K.D. (1999). Mapping Crime: Principle and Practice. National Institute of Justice. Washington DC. 193 pages.

Holland J.H. (1962). Outline for a logical theory of adaptive systems. Journal of the ACM. 9, 297–314.

Holland J.H. (1975). Adaptation in Natural and Artificial Systems, Ann Arbor. University of Michigan Press. 183 pages.

Huang B. and Jiang B. (2001). AVTOP: a full integration of TOPMODEL into GIS. Environmental Modelling and Software. 17(3), 261–268.

Huang B. and Pan X. (2006). GIS coupled with traffic simulation and optimization for incident response. Computers, Environment and Urban Systems. 31(2), 116–132.

Imai A., Sasaki K., Nishimura E. and Papadimitriou S. (2006). Multiobjective simulataneous stowage and load planning for a container ship with container rehandle in yard stacks, European Journal of Operational Research. 171(2), 373–389.

Iori M., S-Gonzalez J.J. and Vigo D. (2007). An exact approach for the vehicle routing problem with two-dimensional loading constraints. Transportation Science. 41, 253–264.

Jankowski P. and Richard L. (1994). Integration of GIS-based suitability analysis and multicriteria evaluation in a spatial decision support system for route selection. Environment and Planning. 21(3), 326–339.

Jha M.K., McCall C. and Schonfeld P. (2001). Using GIS, genetic algorithms, and visualization in highway development. Computer-Aided Civil and Infrastructure Engineering. 16(6), 399–414.

Jiang B. (1996). Cartographic visualization: analytical and communication tools. Cartography. 25, 1–11.

Kantorovich L.V. (1960). Mathematical models of organizing and planning production. Management Science. 6(4), 366–422.

Karimi H.A. and Houston B.H. (1996). Evaluating strategies for integrating environmental models with GIS: current trends and future needs. Computers Environment and Urban Systems. 20(6), 413–425.

Keenan P.B. (1998). Spatial decision support systems for vehicle routing. Decision Support Systems. 22, 65–71.

Kennedy J. and Eberhart R. (1995). Particle swarm optimization In Proceedings of IEEE International Conference on Neural Networks, Perth, Australia. 1942-1948.

Khanafer A., Clautiaux F., El-Ghazali T. (2010). New lower bounds for bin packing problems with conicts. European Journal of Operational Research. 206(2), 281–288.

Kharrat A., Sandu Popa I., Zeitouni K. and Faiz S. (2008). Clustering Algorithm for Network Constraint Trajectory. In: Headway in Spatial Data Handling, Ruas A. and Gold C. (Editors), Lecture Notes in GeoInformation and Cartography, Springer Verlag, Heidelberg. 631–647.

Kolisch R. and Padman R. (2001). An integrated survey of deterministic project scheduling, Omega. 29(3), 249–272.

Korte G. (2000). The GIS Book, OnWord Press, 5th Edition. 400 pages.

Koubarakis M., Sellis T., Frank A.U., Grumbach S., Güting R.H., Jensen C.S., Lorentzos N., Manolopoulos Y., Nardelli E., Pernici B., Schek H.J., Scholl M., Theodoulidis B. and Tryfona N. (2003). Spatio-Temporal Databases: The Chorochronos Approach, Lecture Notes in Computer Science 2520, Springer Verlag, Heidelberg. 352 pages.

Krichen S. and Dahmani N. (2010). A particle swarm optimization approach for the biobjective load balancing problem, Electronic Notes in Discrete Mathematics. 36, 751–758.

Krichen S., Abdelkhalek O. and Guitouni A. (2012a). A bi-objective location area planning for wireless phone network. International Journal of Applied Decision Sciences. 5(4).

Krichen S., Masri H. and Guitouni A. (2012b). Adjacency based method for generating maximal efficient faces in multiobjective linear programming. Applied Mathematical Modelling. In press.

Laporte G. (1992). The traveling salesman problem: an overview of exact and approximate algorithms. European Journal of Operational Research. 59, 231–247.

Laporte G. and Osman I.H. (1995). Routing problems: A bibliography. Annals of Operations Research. 61, 227–262.

Laporte G., Gendreau M., Potvin J.Y. and Semet F. (2000). Classic and Modern Heuristics for the Vehicle Routing Problem. International Transactions in Operational Research. 7, 285–300.

Laurini R. and Thompson D. (1992), Fundamentals of Spatial Information Systems, The APIC series, Academic Press, Etats-Unis. 680 pages.

Laurini R. and Milleret-Raffort F. (1993). Structuring Intelligent Territorial Databases. First International Seminar on Intelligent Systems for Urban Planning. Cagliari, Italy.

Li X. (2011). Emergence of bottom-up models as a tool for landscape simulation and planning. Landscape and Urban Planning. 100, 393–395.

Li Y., Grainger A., Hesley Z., Hofstad O., Sankhayan P.L., Diallo O. and OKtingati A. (2004). Using GIS techniques to evaluate community sustainability in open forestlands in sub-saharan Africa. In Methodologies, models and instruments for rural and urban development. Dixon-Gough. Ashgate Publishing, Aldershot. 146–163.

Liu D.S., Tan K.C., Huang S.Y., Goh C.K. and Ho W.K. (2008). On solving multiobjective bin packing problems using evolutionary particle swarm optimization. European Journal of Operational Research. 190(2), 357–382.

Lodi A., Martello S. and Vigo D. (2002). Models and bounds for two-dimensional level packing problems. Journal of Combinatorial Optimization. 8, 363–379.

Loh K.H., Golden B. and Wasil E. (2008). Solving the one-dimensional bin packing problem with a weight annealing heuristic. Computers and Operations Research. 35(7), 2283–2291.

Lopes R.B., Barreto S., Ferreira C. and Santos B.S. (2008). A decision-support tool for a capacitated location-routing problem. Decision Supply System. 46(1), 366–375.

MacDonald M. (1996). Solid Waste Management models. Journal of solid waste technology and management. 23(2), 73–83.

Maeda S., Kawachi T., Unami K., Takeuchi J. and Ichion E. (2010). Controlling wasteloads from points and nonpoint source to iver system by GIS-aided Epsilon Robust Optimization model. Hydrology-environment Research. 4, 27–36.

Malczewski J. (2006). Gis-based multicriteria decision analysis: a survey of the literature. International Journal of Geographical Information Science. 20(7), 703–726.

Marinakis Y., Marinaki M. and Dounias G. (2010). A hybrid particle swarm optimization algorithm for the vehicle routing problem. Engineering Applications of Artificial Intelligence. 23(4), 463–472.

Marinakis Y. (2012). Multiple phase neighborhood search-grasp for the capacitated vehicle routing problem. Expert Systems with Applications. 39(8), 6807–6815.

Martello S. and Toth P. (1988). A new algorithm for the 0-1 knapsack problem. Management Science. 34(5), 633–644.

Martello S. and Toth P. (1990). Knapsack problems: algorithms and computer implementations, John Wiley & Sons. 296 pages.

Marulli J. and Mallarach J.M. (2004). A GIS methodology for assessing ecological connectivity: application to the Barcelona metropolitan area. Landscape and Urban Planning. 71(2-4), 243–262.

Masri H., Krichen S. and Guitouni A. (2012). Generating efficient faces for multiobjective linear programming problems. International Journal of Applied Decision Science. 15(1), 1–15.

Mendoza J.E., Andres L.M., Nubia V. (2009). An evolutionary-based decision support system for vehicle routing: The case of a public utility. Decision Support Systems. 46, 730–742.

Mitchell P. (1972). Optimal selection of police patrol beats. Journal of Criminal Law. 63, 577–584.

Muttiah R.S., Engel B.A. and Jones D.D. (1996). Waste disposal site selection using GIS-based simulated annealing. Computers and Geosciences. 22(9), 1013–1017.

Moon C., Kim J., Choi G. and Seo Y. (2002). An efficient genetic algorithm for the traveling salesman problem with precedence constraints. European Journal of Operational Research. 140(3), 606–617.

Nemhauser G.L. and Wolsey L.A. (1988), Integer and combinatorial optimization. Wiley. New-York. 763 pages.

Olivera F., and Maidment D. (1999). Geographic information system (GIS) based spatially distributed model for runoff routing. Water Resources Research. 35, 1155–1164.

Parks, B.O. (1993). The need for integration. In: Environmental modeling with GIS, Goodchild M.J., Parks B.O. and Steyaert L.T. (editors). Oxford University Press, Oxford. 31–34.

Perpina C., Alfonso D., Perez-Navarro A., Penalvo E., Vargas C. and Cardenas R. (2009). Methodology based on geographic information systems for biomass logistics and transport optimisation. Renewable Energy. 34(3), 555–565.

Peters B., Smith J. and Venkatesh S. (1996). A control classification of automated guided vehicle systems. International Journal of Industrial Engineering. 3, 29–39.

Phillips P. and Lee I. (2010). Crime analysis through spatial areal aggregated density patterns. Geoinformatica. 15, 49–74.

Pullar D. and Springer D. (2000). Towards integrating GIS and catchment models. Environmental Modeling and Software. 15(5), 451–459.

Rossi F. and Villa N. (2009). Optimizing an organized modularity measure for topographic graph clustering: a deterministic annealing approach. EEG Neurocomputing. 73, 1142–1163.

Rupp T.S. (1996). Landscape-level modeling of spruce seedfall using a geographic information system. Proceedings of 3rd International Conference on Integrating GIS and Environmental Modeling. NCGIA, Santa Barbara.

Ruiz R., Maroto C. and Alcaraz J. (2004). A decision support system for a real vehicle routing problem. European Journal of Operational Research. 153, 593–606.

Ryu K.H. and Ahn Y.A. (2001). Application of Moving Objects and Spatiotemporal Reasoning, A TimeCenter Technical Report TR-58, Aalborg University, Denmark.

Santos L., Coutinho-Rodrigues J. and Antunes C.H. (2011). A web spatial decision support system for vehicle routing using google maps. Decision Support Systems. 51, 1–9.

Scheithauer G. (1992). Algorithm for the container loading problem. Operational Research Proceedings. Springer Verlag, Heidelberg. 445–452.

Schumann A.H., Funke R. and Schultz G.A. (2000). Application of a geographic information system for conceptual rainfall-runoff modeling. Journal of Hydrology. 240, 45–61.

Shao L. and Ehrgott M. (2007). Approximately solving multiobjective linear programs in objective space and an application in radiotherapy treatment planning, Mathematical Methods of Operations Research. 68(2), 257–276.

Shyy T.K., Stimson R. and Chhetri P. (2007). Web-based GIS for mapping voting patterns at the 2004 Australian federal election. Applied GIS. 3(11), 1–20.

Singh V.P. and Fiorentino M. (1996). Geographical information systems in hydrology. Kluwer Academic Publishers, The Netherlands. 443 pages.

Stuart, N. and Stocks, C. (1993). Hydrological modeling within GIS: an integrated approach. Application of Geographic Information System in Hydrology and Water resources. The Netherlands. 211, 319–329.

Sutapa S. and Jha M.K. (2011). Modeling a rail transit alignement considering different objectives. Transportation Research Part A: Policy and Practice. 45(1), 31–45.

Sui D.Z. and Maggio R.C. (1999). Integrating GIS with hydrological modeling: practices, problems, and prospects. Computer, Environmental and Urban Systems. 23, 33–51.

Sumathi V.R., Natesan U. and Sarkar C. (2007). GIS based approach for optimized siting of municipal solid waste landfill. Journal of Waste Management. 28(11), 2146-2160.

Sui D.Z. and Maggio R.C. (1999). Integrating GIS with hydrological modeling: practices, problems and prospects. Computers, Environment and Urban Systems. 23(1), 33–51.

Tait J., Williams R. and Lyall P. (2000). Roadmapping Foresight: Monitoring and Evaluation of Complex Programmes. SUPRA Report to the Office of Science and Technology.

Tait N.G., Davison R.M., Whittaker J.J., Leharne S.A. and Lerner D.N. (2004). Borehole Optimisation System (BOS)—A GIS based risk analysis tool for optimising the use of urban groundwater. Environmental modelling Software. 19(12), 1111–1124.

Tarantilis C.D. and Kiranoudis C.T. (2002). Distribution of fresh meat. Journal of Food Engineering. 51(1), 85–91.

Tarantilis C.D., Diakoulaki D. and Kiranoudis C.T. (2001). Combination of geographical information system and efficient routing algorithms for real life distribution operations. European Journal of Operational Research. 152(2), 437–453.

Teghem J., Tuyttens D. and Ulungu E. (2000). An interactive heuristic method for multi-objective combinatorial optimization. Computers and Operations Research. 27, 621–634.

Tomlinson R.F. (1984). Geographic Information Systems—A New Frontier. The Operational Geographer. 5, 31–35.

Topaloglu H. (2006). A parallelizable dynamic fleet management model with random travel times. European Journal of Operational Research. 175, 782–805.

Toth P. and Vigo D. (2003). The granular tabu search and its application to the vehicle routing problem. Informs Journal on Computing. 15, 333–348.

Tuyttens D., Teghem J., Fortemps Ph. and Van Nieuwenhuyze K. (2000). Performance of the MOSA method for the bicriteria assignment problem. Journal of Heuristics. 6(3), 295–310.

Ulungu E. and Teghem J. (1994). Multi-objective combinatorial optimization: a survey. Journal of Multi-Criteria Decision Analysis. 3, 83–104.

Vairavamoorthy K., Gorantiwar S.D., Yan J. and Galgale H.M. (2006). Water Safety Plans: IRA-WDS Software and Manual for Risk Assessment of Contaminant Intrusion into Water Distribution Systems. WEDC. 124 pages.

Vairavamoorthy K., Yan J., Galgale H.M. and Gorantiwar S.D. (2007). IRA-WDS: a GIS-based risk analysis tool for water distribution systems. Environmental Modelling Software. 22(7), 951–965.

Vigo D. (1996). A heuristic algorithm for the asymmetric capacitated vehicle routing problem. European Journal of Operational Research. 89, 108–126.

Wang X., Yu S. and Huang G.H. (2004). Land allocation based on integrated GIS-optimization modeling at a watershed level. Landscape and Urban Planning. 66(2), 61–74.

Waugh T.C. (1986). The Geolink System, interfacing large systems. Proceedings of Auto Carto, London. 76–85.

Wascher G., Hausner H. and Schumann H., (2007). An improved typology of cutting and packing problems. European Journal Of Operational Research. 183(3), 1109–1130.

Wu F. (1998). SimLand: a prototype to simulate land conversion through the integrated GIS and CA with AHP-derived. International Journal of Geographical Information Science. 12(1), 63–82.

Xie R. and Shibasaki R. (2005). A unified spatiotemporal schema for representing and querying moving features. ACM SIGMOD. 34(1), 45–50.

Zachariadis E.E., Tarantilis C.D. and Kiranoudis C.T. (2009). A guided tabu search for the vehicle routing problem with two-dimensional loading constraints. European Journal of Operational Research. 195, 729–743.

Zbidi N., Faiz S. and Limam M. (2006). On mining summaries by objective measures of interestingness. Machine Learning Journal, Kluwer Academic Publishers. 62(3), 175–198.

Zhang Z. and Grith D.A. (1997). Developing user-friendly spatial statistical analysis modules for GIS: an example using ArcView. Computers, Environment and Urban Systems. 21(1), 5–29.

Zhang Y. Jinpeng L.V. and Ying Q. (2010). Traffic assignment considering air quality. Transportation Research Part D: Transport and Environment. 15(8), 497–502.

Zhu Z., Huang W. and Lim A. (2011). A prototype column generation strategy for the multiple container loading problem. 9th Metaheuristics International Conference (MIC'2011), Udine, Italy.

Index

Color Plate Section

Introduction

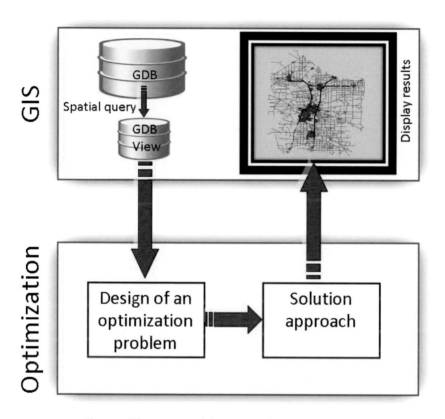

Figure 1 The process of decision making inside a GIS.

Chapter 1

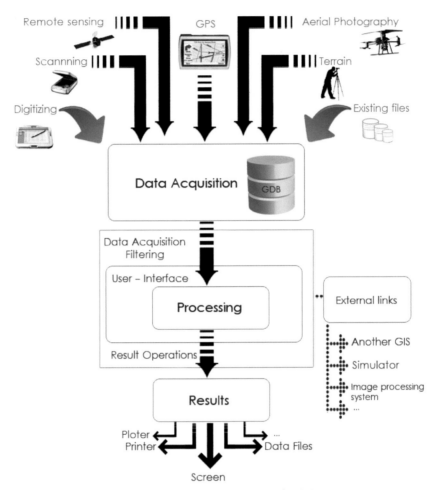

Figure 1.4 Main components of a GIS.

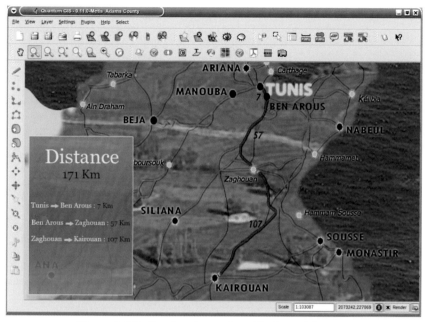

Figure 1.5 Optimal pathway from Tunis to Kairouan.

Chapter 2

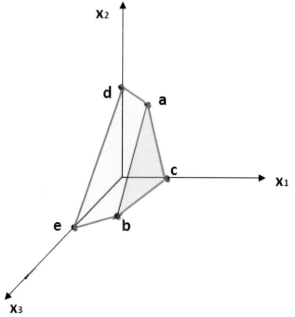

Figure 2.2 Feasible space of an optimization problem.

Chapter 4

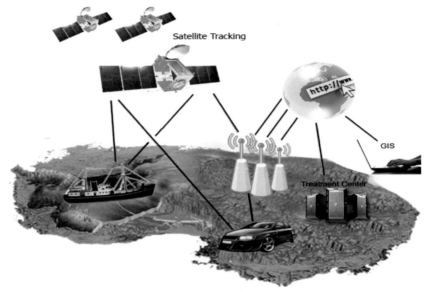

Figure 4.1 The geotracking process.

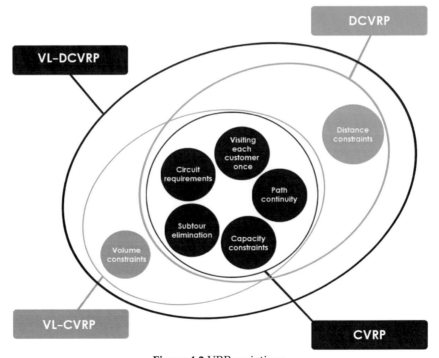

Figure 4.2 VRP variations.

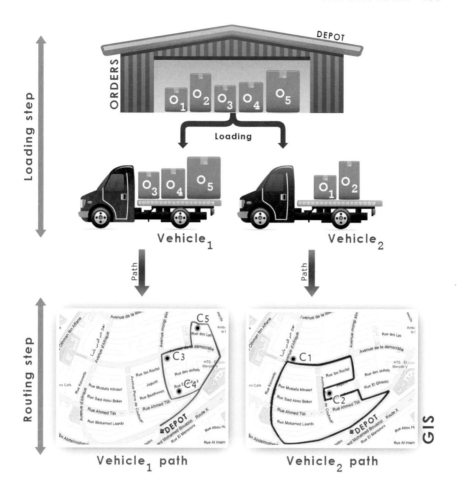

Figure 4.3 The VL-DCVRP process.

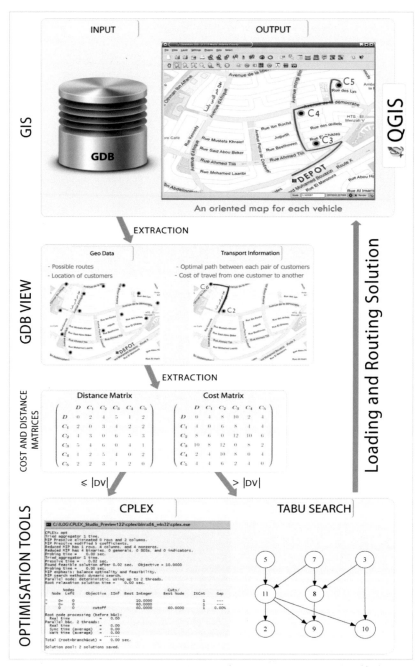

Figure 4.7 A GIS-O loose integration of QGIS and CPLEX-TS.

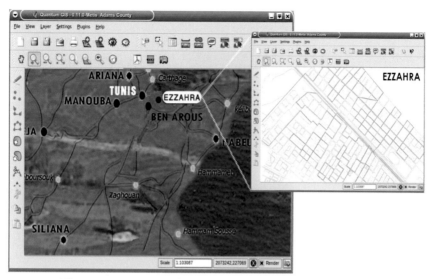

Figure 4.15 Location of the area under study: Ezzahra town.

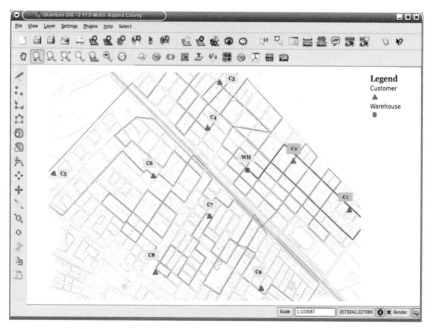

Figure 4.16 The routing of vehicle 1 for $m = 9$ and $n = 3$.

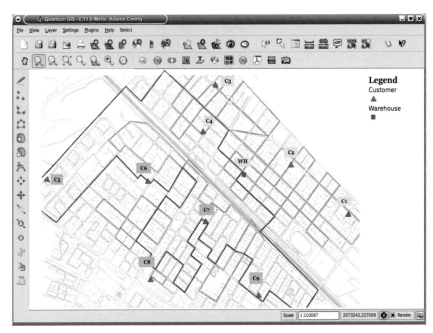

Figure 4.17 The routing of vehicle 2 for $m = 9$ and $n = 3$.

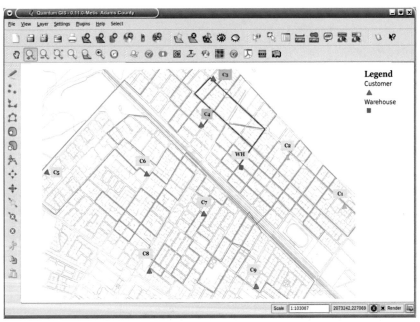

Figure 4.18 The routing of vehicle 3 for $m = 9$ and $n = 3$.

For Product Safety Concerns and Information please contact our EU representative GPSR@taylorandfrancis.com Taylor & Francis Verlag GmbH, Kaufingerstraße 24, 80331 München, Germany

Printed and bound by CPI Group (UK) Ltd, Croydon, CR0 4YY

01/05/2025

01858615-0001